Maritime Simulation

Proceedings of the
First Intercontinental Symposium
Munich, June 1985

Edited by Moshe R. Heller, CONTROL DATA

With 104 Figures

Springer-Verlag
Berlin Heidelberg NewYork Tokyo 1985

First Intercontinental Maritime Simulation Symposium and
Mathematical Modelling Workshop, Munich June 3-5, 1985

Initiated, sponsored and organized by:
Control Data GmbH, Stresemann-Allee 30, 6000 Frankfurt, FRG

Co-Sponsors:
The International Marine Simulator Forum (IMSF)
The Society for Computer Simulation (SCS)

Editor:
Moshe R. Heller, Control Data GmbH, FRG

ISBN-13:978-3-642-82562-0 e-ISBN-13:978-3-642-82560-6
DOI: 10.1007/978-3-642-82560-6

2161/3020-543210

Preface

Welcome to Bavaria - Germany and to the First Intercontinental Maritime Simulation Symposium and Mathematical Modelling Workshop. A triennial international conference jointly promoted by Control Data, IMSF and SCS, which takes place at Schliersee, a small town near the Alps.

The aim of the Symposium is to cover most of the aspects of maritime modelling and simulation in theory and practice, to promote the exchange of knowledge and experience between different international research groups in this field, and to strengthen the international contact between developers and users of modelling and simulation techniques.

On the occasion of the Symposium people of scientific and engineering disciplines will meet to discuss the state-of-the-art and future activities and developments.

A large number of contributed papers has been strictly examined and selected by the papers committee to guarantee a high international standard.

The book contains the accepted papers which will be presented at the Symposium. The papers have been classified according to the following topics:

1. Fifth Generation Computer Technology
2. Simulation-Software-Tools
3. An Industrial Computer System - The Chrysler Story
4. Marine Mathematical Modelling
5. CFD for Marine Vehicles
6. Navigation Methodology
7. Marine Maneuvering and Motion Simulation
8. Off-Shore Modelling
9. Steering and Control of Marine Vehicles
10. Training and Traffic Control
11. Under-Water Vehicles Operation

Authors from 9 countries will meet at the Symposium. They work for Industrial Companies, Universities and the Military so that a broad spectrum of simulation activities is covered: Theory and application, hardware and software, research and operations.

The editor is greatful to the authors for making possible the publication of this book, and especially to the papers committee J. B. VAN DEN BRUG, STEVE CHISLETT, IAN R. MCCALLUM and JOSEPH J. PUGLISI, for the selection of the papers and the contribution to the success of the Symposium.

Mr. VON HAGEN and Ms. RAUFELDER of Springer-Verlag for the excellent publication of the proceedings to whom I would like to extend my thanks.

My thanks also go to all of the Control Data people who have been involved beyond everyday's work in the promotion of the Symposium.

Munich, June 1985 M. R. Heller

Contents

Session 4:

Marine Mathematical Modelling

Session 5:

CFD for Marine Vehicles

Session 6:

Navigation Methodology

Session 7:

Marine Maneuvering and Motion Simulation

Introduction
Simulation »In the Boat«

M. R. HELLER

Control Data GmbH, München, W-Germany

Before the widespread use of the digital machines, relationships and activities in the scientific-engineering approach were largely determined by the human mind, its formalization and analytic powers. The scientist or engineer who approaches a real world process tries to gain insight or an understanding of the phenomena on the process under study. One of the very powerful methods consists in trying to obtain an abstract or formal model or representation of the process. The activity is defined as m o d e l b u i l d i n g and f o r m a l i z a t i o n. In essence the procedure requires abstraction and simplification. Simplification is necessary to restrict the complexity of the representation. One only chooses those properties within given boundaries of space or time which are believed to be connected with each other but unconnected with other properties or other parts of the world. Basically the model builder proceeds by hypothesis, induction and deduction. The complete body of methods is called modelling methodology.

Modelling itself is for a part still considered as an art. There are a large number of factors that come into play. The representation process involves almost always "inter- or extrapolation". Certainly, a large body of methodology is objective and mathematically sound. Many tools are well-defined and clearly stated. Statistical techniques and parameter estimation procedures have a firm logical basis, but standard techniques cannot solve all problems. The final product is the result of careful trade-off's between existing facts, decision on the choice of representative details, careful experimental work and its interpretation. Examples of such issues relate to the choice of formalisms, the evaluation of the validity of à priori facts, the required level of descriptive details etc.

> Ancient Greeks had problems with deception. The legend tells us that Ulysses was tempted at sea by the sirens who allured him into destruction with their treacherous songs and it was only by training his crew to tie him to the mast that he could avoid the ruin.

In modern electronic world it is JUST AS NECESSARY TO HAVE A HIGHLY TRAINED CREW not only for ONE anticipated mode of deception, but for a horde of possible situations.

It has sometimes been said that systems with automatic support devices are far better off than a system with a crew that has to evaluate threat and then take precautions; the automation is thought to be faster and without having THE TROUBLE WITH HESITATION BETWEEN STEPS to be taken.

M o d e l   i n t e g r a t i o n is introduced; different descriptions of the same real world process are compared, screened for consistency and integrated into a whole. This activity is a basic step in scientific work, especially in the process of developing a theory and working out general principles. The formal model, though a simplified representation of reality, always summarizes a vast amount of information, comprising facts, axioms and hypotheses. If its validity is high or in

ULYSSES AND SIRENS

Museum Dahlem, BERLIN

other words if its descriptive quality is good, it can be used to obtain useful knowledge on the system under study. In this sense a model can be seen as an extremely compact and useful extension of a data base containing loose pieces of data. With modern computerized receivers and possibly computercontrolled system arrays A HIGHLY EDUCATED OPERATOR CAN LOOK THROUGH DENSE ENVIRONMENT and give decision-makers correct information.

The s i m u l a t i o n a c t i v i t y, being the experimentation with models, supports not only the model utilization but the model building as well. The introduction of computational devices requires provisions for suitable man-machine communication. The nature of this communication has its impact on the modelling methodology. Man is especially apt in reasoning and in recognizing patterns; his computational powers however are limited. A process modelbase is built up, composed of several candidate models, "primitives" and combined ones, with their own features. The creation and testing of an extensive variety of such models is of key importance and relies upon an efficient use of advanced parallel processing. At present there are arguments that, with time, modelling methodology will be incorporated for its major part in "intelligent", "self-organizing" machines, so that human intervention could be brought to a bare small level. In that case simulation will in a sense supplant modelling. More over it can be stated that SIMULATION is a COST EFFECTIVE solution to engineering effectiveness.

Nowadays in advanced information processing a fundamental transition is taking place from data-processing to k n o w - l e d g e - p r o c e s s i n g, which will be the key of the next-generation computers. Advanced information processing uses the automation of data acquisition and data processing as well as the automation of the reasoning process in order to combine these functions to create a system capable of showing an "intelligent" behaviour. This will allow simulators to rely on flexible, human-like thought processes to diagnose problems, rather than on rigid procedures expressed in flowcharts or "decision trees". To incorporate this approach , the central knowledge-base contains rules generated from discussions with the specialist of the process under study. This added flexibility will greatly improve the diagnostic ability of current simulators. New simulator architectures enable these developments in expert system design.

The ultimate goal in process-studies consists in the integration of knowledge from the model level to the à priori level. Therefore, existing models must be generalized. Extensive experimentation, results in the validation of the generalized models; and last but not least these general models have to be interpreted in laws and theories. Future developments should reveal these frontiers of simulation.

References

(1) Van Steenkiste, G.C.: Simulation "On the Road", The First European Cars/Trucks Simulation Symposium, Schliersee, May 2-4, 1985

(2) Svensson, G.: Use of Airborn Jammers for ECCM Training of Operators, EW Conference, London Dec. 8-9, 1983

Session 1 Fifth Generation Computer Technology

The Introduction to Parallel Processing to Solve Future Applications

WAYNE A. RAY

Director Parallel Processing
Control Data Corporation
Minneapolis, Minnesota, U.S.A.

Parallel Processing is an old methodology, and is now becoming part of the design of new applications. The evolution of VLSI technology and the parallel processing software has provided a base technology for commercial organizations to introduce high parallel processing into the marketplace.

Parallel Processing is not a revolutionary concept, but a very simple fundamental way of doing computations. Parallel Processing is defined as the ability for the application user to have more than one processor working on the same job task at the same time. In some communities it is called multi processing, in other communities it is called parallel processing and in other communities it is called multi parallel processing. One of the problem that has confronted application designers over the last N years is the inability of the computer manufacturer to deliver the performance of computing systems to match the needs of the scientists. Let us look at the history of computing power.

During the last 30 years, there has been a number of significant changes of capability and each change initiated a need to improve or extend the computer technology. As an example, the first computer was designed to generate tables for scientists. Once completed, it became evident that the computer could be used to solve the real problem and the need for tables was no longer the goal of scientists.

During the last 30 years, the distinct change of capability were:

GENERATION OF TABLES FOR SCIENTISTS
ASSEMBLERS FOR DIGITAL COMPUTERS
HIGH LEVEL LANGUAGES (FORTRAN)
MULTIPROGRAMMING COMPUTER SYSTEMS
INTERACTIVE JOB PROCESSING
SUPERCOMPUTERS - VECTOR PROCESSING
PERSONAL COMPUTER - WORK STATIONS

Each of the above provided the application designer, significant tools to solve larger problems yet each of the capabilities forced the application designer to subset the problem, change the mission to fit the computer architecture or change the science to accomodate the computer system limitations.

With the additional capability, the size and scope of the application task also increased to exceed the performance capability of the computer system. If the new computer increased the performance of the application by a factor or 4, within weeks, the scientist increased the scope by a factor greater than 4.

The 6600 introduced by Control Data in 1962 set a new standard for computer performance systems that are now called Supercomputers. The term Supercomputer is a relatively new term in the computer industry but in reality the Supercomputer is the fastest machine known in any instance of time. No matter how powerful it is, it is still one order of magnitude less capable than what is required by the applications user. The 6600 broke the computation barrier where in the early 50's the speed of the machine had been reduced to a cycle time of 2 micro-seconds. Each computation required two or more machine cycles to execute an instruction. The Control Data 6600 improved the performance of the machine by a factor of 20 and its cycle time was reduced to 100 nanoseconds.

During the next 20 years, from 1964 to the present time, we have seen the cycle time of machine go down. The following chart addresses the performance of the machines based upon machine cycle times.

IBM 7094	2000	nanoseconds
CDC 6600	100	nanoseconds
CDC 7600	27.5	nanoseconds
CDC 205	20	nanoseconds
CDC 990	16	nanoseconds
CRI XMP	9.5	nanoseconds

Hardware designers have been able to reduce the number of cycles to perform an instruction and overlap instructions to further increase the performance of a processor.

So in the last twenty years, we have seen the machines power increase by only a factor of 10 yet the applications require the computer performance to improve by a factor of 100 or a factor of 1000 to handle the total application that is needed to be solved.

The last three Computer Systems integrate vector hardware that greatly increased the potential compute power of their system. Unfortunately, most applications can not use the vector hardware 100 % of the time. In order to better utilize the vector performance, the vector computers require the applications designer or programmer to change the code or the algorithm. Vectorizing Fortran compilers have made significant progress but the percent of vectorizing still a long way from a 90 % vectorization of application codes.

If we look to the future of hardware components, the market is seeing machines with a cycle time of 6 nano-seconds, 4 nanoseconds, but nobody has really said we will go into the picoseconds. If we continue down the path of performance based upon pure cycle time, we would need a computer with a machine cycle time of 50 to 100 picoseconds in order solve the problem of the future. The goal of most component engineering is to be able to deliver a 1 nano-second machine before the end of the 1980's. That is still is factor of 10 to 20 times slower then the needs of the user.

How will Parallel Processing address the user requirement? Parallel Processing is an old technology and old methodology.
The first computers were Parallel Processing machines, but did not have the tools or the software to allow the user to use and exploit the parallel capability of the hardware. Thus computers went to the serial, serial instruction, the Von Newmann type architecture. The problem with Parallel Processing is the lack of tools, the lack of systems that will allow the user to develop the decomposition of application code for a Parallel Processing System. What is needed is the ability to link together in a very efficient manner a number of processors to do the total application.

The problem facing the Parallel Processing developer is what is the granulatory required by applications. Most computer scientists, feel that there is only of granulatory and that is at the hardware level, but there is a second measure of granulatory that is the applications and what is nutural to the application. If the granulatory, or the hardware performance is low that application will need to break the code, the application, the algorthim into thousands of processors. The application may not contain adequate breaks in the computational structure to achieve a 95 % degree of parallelism in order to achieve the performance level needed by the application. The ability to stay on the linear curve of the performance is almost impossible for large number of processors (100). Many theoretical examples have shown that a thousand processor system executing a task with a 90 % level of parallelism would provide the performance of a five processor system. This is less than .5 % of the total peak performance of of the parallel processor system.

A Parallel Processing System containing high performance processors (the high level granulatory) requires an interconnection architecture that allows the application to transmit data at the speed of an individual processor in the system.

A Parallel Processing System requires at least two of the following capabilitites in order for the applications to use the parallel hardware:

1. High Level Languages
2. Software to control execution and synchronization
3. Tools to measure parallel bottlenecks
4. Parallel library routines
5. Algorithims for parallel structures

The high level language requirement is the ability to take the present applications code and have the compiler automatically generate the parallel code to be run in the parallel processing system. The problem to date is that the largest percentage of code in existance is in Fortran. This language does not have the advanced constructs that are needed to do effective parallel processing. Pascal and Ada are examples of languages that would allow for effective utilization of a parallel operating environment.

The software that is required to do parallel processing must provide a high level ease of use that would allow the application to have control over individual processing elements within the system. The user needs to be able to initiate a task, status the task and have software to provide the necessary task synchronization to support the data transfer between processors in the system. Software is also needed to provide path and trace flow since the problem of debugging complex parallel systems is an order of magnitude more difficult than the present application verification process.

In the area of tools, the parallel processing application developer will need a powerful performance measurement and analysis capability in order to determine the bottlenecks in a parallel application.

An application will require the same library capability that presently exist on conventional machines. This parallel library must contain a dynamic capability then will expand to use all of the resources that are assigned to a given application.

The last software ist the re-evaluation of the present computer algorthims and a development of a parallel algorthim to perform the computational structures. Initial evaluation has indicated that a parallel algorithm could provide an order of magnitude improvement in processing speed compared to the present uni-processor algorithm.

SUMMARY

Parallel Processing offers the applications user the ability to achieve a quantum jump in the size and scope of applications. Parallel Processing Systems are appearing in the marketplace in an increasing number and will provide experimental systems for computer scientists to develop the parallel processing tools for parallel processing applications.

Supercomputers and Mathematical Modeling

G. V. Tarsy,

ETA Systems, Inc.
St. Paul, Minnesota

ABSTRACT

The rise of supercomputers and the role in mathematical modeling is discussed with examples from error dynamics, electronic circuits and numerical linear algebra.

The supercomputer has been variously defined as:

1. A computer with the fastest processor able to address the most memory.
2. A computer with the capability only one order of magnitude less than current scientific/technological requirements.
3. A computer that cost ten million dollars.

Though definition 2 may be whimsical, and definition 3 perhaps facetious, a glance at figure 1 indicates that definitions 1 and 3 seem to jibe at least in recent times. That definition two is also valid remains to be seen. In any case definition 1 implies that supercomputers have always existed although the term is of recent coinage.

The mainstream of supercomputer development is often thought to have begun with the CDC 6600 Computer. Its later history has progressed as given by the chronological sequence:

CDC 6600
CDC 7600/IBM 360-195
TIASC/CDC Star 100
Cray 1
CDC CYBER 205
Cray XMP/Fujitsu VP400/Hitachi S810/20

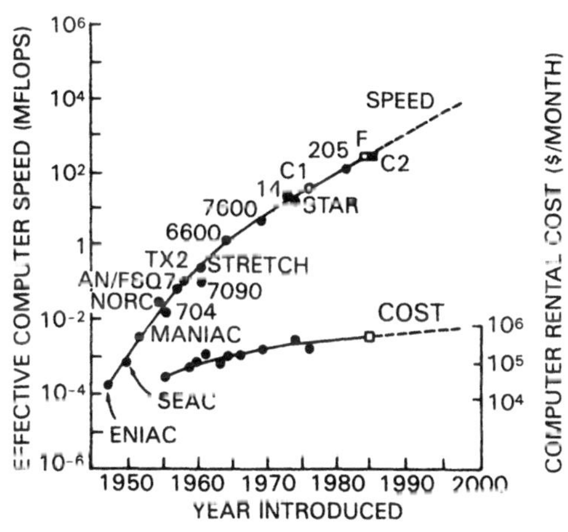

Figure 1. Past Growth and Future Projections for Computer Speed and Computer Cost.
Source: "The Influence of Computational Fluid Dynamics on Experimental Aerospace Facilities", National Academy of Sciences, Washington, D.C., 1983.

This progression corresponds to an architectural evolution that can be characterized by the chain:

overlapped from the von Neumann processing (pipeline SISD)

↓

vector processing (SIMD)

↓

overlapped vector processing (pipeline SIMD)

↓

multiprocessor processing (MIMD)

where the acronyms SISD, SIMD, and MIMD denote respectively single instruction stream, single data stream, single instruction stream, multiple data stream, multiple instruction stream, multiple data stream.

It is thus clear that recent and current emphasis on supercomputer architecture has been on vector processing which may be characterized as a form of parallelism afforded by implementing segmented operations on arrays of quantities. We shall discuss this in greater detail later.

The evolution of supercomputer architecture can be understood better in the light of the fact there are two main ways to speed up a computer.

1. Make basic indivisible/operations faster.
2. Do more operations simultaneously.

The first way is primarily a function of technology, the second of architecture.

Since the desired order-of-magnitude speed increase for supercomputer generation is no longer achievable solely by technology, parallelism has been the main method of acceleration in recent times and will be for the near future. These facts are expanded upon and organized in tables 1 and 2.

TABLE 1. CURRENT SUPERCOMPUTERS

	FUJITSU VP-200	HITACHI S-810/20	CRAY 1M	CRAY X-MP	CDC 205
First Installation	Nov. 1983	Nov. 1983	Sep. 1983	June 1983	Feb. 1982
Architecture (all 64-bit words)	Vector, IBM compatible	Vector, IBM compatible	Vector	Vector multi-processor	Vector
Performance					
Peak (Mflops)	500	630	250	630	400
Cycle time (ns)	15 (vector) 7.5 (scalar)	15	12.5	9.5	20
Max. Main Memory Size (millions of bytes)	256	256	32	32	64
Technology					
Logic (gate delay)	ECL (350 ps)	ECL (350 ps)	ECL (SSI, 0.7 ns)	ECL (0.5-1 ns)	ECL (700 ps)
Main memory	64K MOS Static (55 ns)	16K MOS Static (40 ns)	16K MOS (70 ns)	4K ECL (25 ns)	MOS (45 ns)
Input/Output					
No. paths (total)	32	32	19	19	16
Max. data rate path (millions of bytes/sec.)	3	3	100	1,000	25
Max. aggregate data rate (millions of bytes/sec.)	900	1,100	550	1,550	400
Cooling	Air	Water (vector unit)	Freon	Freon	Freon

Source: Riganati and Schneck, Supercomputing, Computer 10/74

TABLE 2. SUPERCOMPUTERS IN DESIGN

	CRAY 2	CRAY 3	ETA GF10	DENELCOR HEP-2	NEC SX-2
Expected Installation	Late 1984	1986	1986	1986	1985
Architecture (all 64-bit words)	Vector multipro- cessor		Vector multipro- cessor	Scalar	Vector
Performance					
Peak (Mflops)	1,000*	NA	10,000	4,000	1,300
Cycle time (ns)	4	1	NA	12.5	6
Max. Main Memory Size (millions of bytes)	NA	NA	256	4,000	256
Technology					
Logic (gate delay)	ECL	ECL GaAs	CMOS (0.5-1 ns)	ECL	NA
Main memory	MOS	MOS	MOS	MOS	64K MOS Static
Input/Output					
No. paths (total)	NA	NA	16	NA	32
Max. data rate path (millions of bytes/sec.)	NA	NA	50-60		1.5
Max. aggregate data rate (millions of bytes/sec.)	NA	NA	1,000	NA	1,350
Cooling	Liquid emersion	NA	Nitrogen (100°K) Dehydrated chilled air (memory)	Air	Water

*Derived from 6-12 times. NA = not available. Vector/options multiprocessor.

Source: Ibid Table 1.

The first 15 years of supercomputing — (1965-1980 assuming we start with the CDC 6600) new supercomputers were mostly restricted to government laboratories, government funded research, educational institutions, and the few industrial companies doing structural design and oil exploration. Ten years ago Cray Research predicted a world market for the Cray 1 of not more than two dozen machines. Today, however, there are almost 100 supercomputers installed worldwide (about 65 Crays, and 25 CYBER 205s) and recent forecasts call for 400 to 1000 customers for supercomputers by 1990. What has accounted for the recent growth of Super Computing and the current predictions?

Although there are several facts which account for the increase, perhaps the most significant one is that supercomputers seem to have reached the threshold in performance. In many areas of science and technology, computer simulations are now possible which approach or surpass the efficacy of older methods of research and development at a sharply reduced cost compared to those methods. The growth of the commercial supercomputer market has been viewed in these terms.

> "The commercial supercomputer market will take off as manufacturers begin to see significant possibilities in non-defense applications. The big potential is among automotive, aerospace, chemical, semi-conductor, and biomedical suppliers."
>
> Leo Kramer, VP Operations
> ETA Systems

> "The role of supercomputers in simulation will profoundly affect the competitiveness of U.S. industry. By revolutionizing the way engineers design and test products, supercomputers will accelerate development cycles and improve offerings."
>
> Peter A. Gregory, V.P.
> Corporate Planning, Cray Research
> "High Technology" (May 1984)

Let us consider two areas where supercomputer modeling is significant; aeronautical engineering and electronic circuit design. The following discussion of aircraft design is based on a presentation given by W.F.Walhouse, Director of NASA AMES Research Center, at the Conference "Supercomputers", 2/84, sponsored by the DPMA. The numerical aerodynamics simulation (NAS System) to be developed by NASA was described.

In the "distant" past, design of aerodynamic configurations was principally based on experiment - mostly in wind tunnels. More recently, computer modeling has come to play a complimentary role. The respective areas of contributions can be summarized as follows:

Experiment:

—detailed performance data once final configurations are built, complex configurations,

—complex aerodynamic phenomena,

—computer code calibration.

Computation:

—flow physics,

—new design concepts,

—configuration optimization,

—instrumentation of experimental models,

—test facility error corrections,

—data for conditions beyond experimental operating range.

Both techniques are subject to simulation errors:

Experiment:

—wall and support interference,

—model distortion,

—operating range limitations,

—flow nonuniformity.

Computation:

—phenomena logical errors,

—numerical errors,

—limitations on computer speed and capacity.

The last item is at the crux of the supercomputer development rationale. By continuing the use and development of supercomputers, the problems summarized in figure 2 could be avoided.

- Best available methods were inadequate to uncover design problems before flight test.
- Some problems could have been avoided using **today's** computational capability.
- Others would require **NAS Level** capability.

Aircraft	Problems Discovered in Flight Test	Consequence
C-141	Incorrectly Predicted Wing Flow	Compromised Performance, Costly Modifications
C-5A	Incorrectly Predicted Drag-Rise Mach Number	Reduced Wing Fatigue Life
F-111	Incorrectly Predicted Transonic Airframe Drag	Costly Modifications
B-58 B-70 YF-12	Incorrectly Predicted Transonic Performance	Reduced Aircraft Effectiveness
F-102 F-106	Incorrectly Predicted Transonic Drag	Reduced Performance
2 Civil Transports	Incorrectly Predicted Nacelle-Wing Interference	Redesign Required

Figure 2.

An example of the savings available through computer simulation is diagramed in figure 3 which pertains to the highly maneuverable aircraft technology remote piloted research vehicle of Rockwell International Corporation.

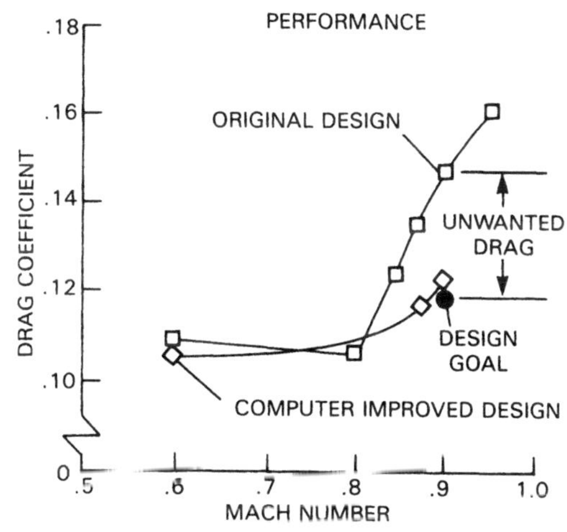

Figure 3.

It is unlikely that the design goal would have been reached within budget without computer simulation. Moreover, the cost of redesign by computer was just 4 percent of using conventional methods.

Another example is the increased fuel efficiency of the air bus A310, principally achieved by computational aerodynamics. The A310 is 20 percent more fuel efficient than the A300 which projects to a potential savings of 10 billion dollars over the 15-year life of a 400 airplane fleet.

These achievements have come through the ability to solve specific sets of governing equations on supercomputers. Table 3 shows how these methods relate to supercomputing for now and the future. It is easily seen that computational aerodynamics will absorb any available supercomputer capability for the foreseeable future.

Table 3. Governing Equations and Computer Requirements for Computational Aerodynamics

Approximation	Capability	Grid Points Required	Computer Requirement
Linearized Inviscid	Supersonic/Supersonic Pressure Loads Vortex Drag	$3 = 10^3$ Panels	1/10 Class VI
Nonlinear Inviscid	Above Plus: Transonic Pressure Loads Wave Drag	10^5	Class VI
Reynolds Averaged Navier-Stokes	Above Plus: Separation/Reattachment Stall/Buffet/Flutter Total Drag	10^7	30 x Class VI
Large Eddy Simulation	Above Plus: Turbulence Structure Aerodynamic Noise	10^9	3000 x Class
Full Navier-Stokes	Above Plus: Laminar/Turbulent Transition Turbulence Dissipation	10^{12} to 10^{15}	3 Million to 3 Billion Class VI

Class VI = 50 Million Floating Point Operations/Second (Mflops)

Our second example comes from the practices of the integrated circuit manufacturer. A basic problem for the circuit designer is that the new VLSI systems are so complicated that dealing with the complexity alone overwhelms all other problems. Figure 4 outlines the features of design automation.

The most computationally intensive module of design automation is circuit simulation which absorbs between 40 and 60 percent of CPU power over various design regimes and all computer modules.

A typical circuit simulator views a circuit as a graph whose nodes are junctions for the edges which represent circuit components, semiconductor devices, resistors, capacitors, etc.

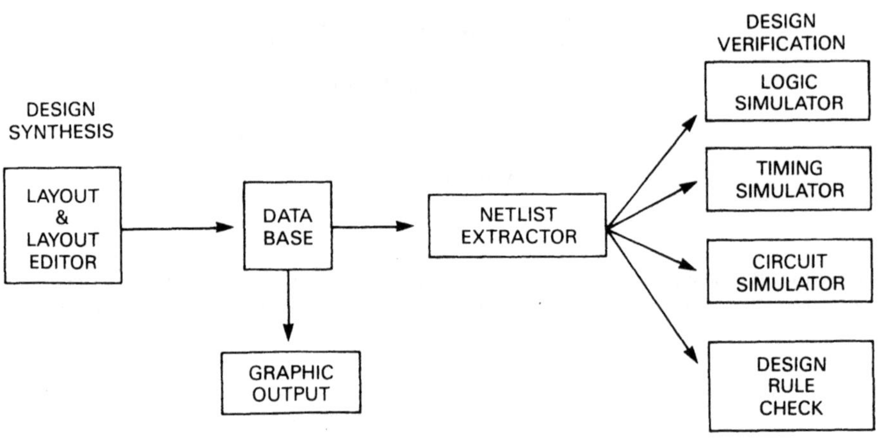

Figure 4. Design Automation

The behavior of the circuit, principally current flow, is described by a set of nonlinear differential equations. The solution of these equations is discretized over time and solved as diagramed in figure 5.

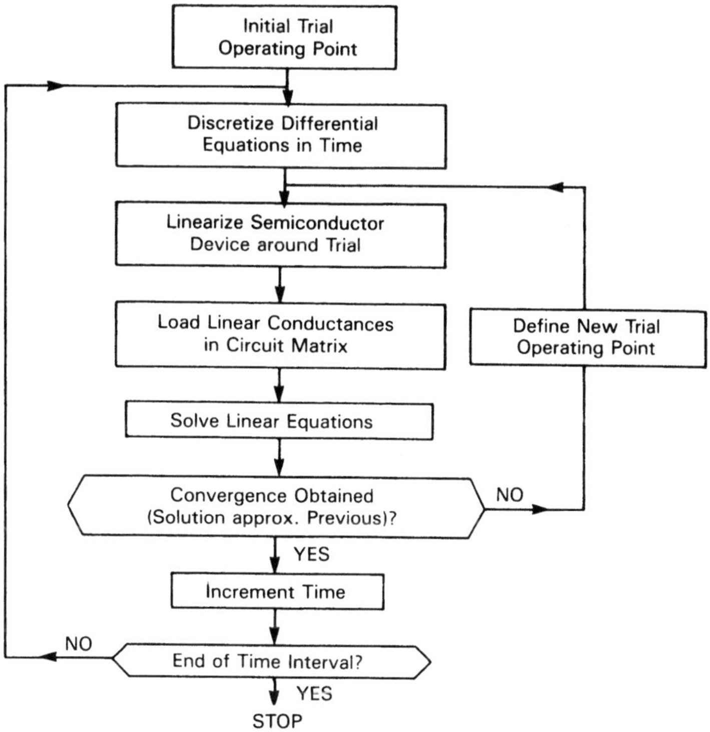

Figure 5.

The two major tasks in this simulation are device evaluation and linear equation solution. However, the equation solution, because of its exponential computational complexity as a function of the number of devices, is the dominating task for large circuits. For example, in a 3000 device circuit, the equation solution accounts for 90 percent of the computational load.

For a circuit of this size, supercomputer circuit simulators outperform an IBM 3081C by factors of 10 - 20 and a VAX 11/780 by factors of 120 to 300. Current and future design requirements are for much larger circuits than these with attendant greater advantages to be obtained from supercomputer models.

If this were the best of all possible worlds then the price of a supercomputer would be all that one would need to pay to obtain the previously discussed benefits. However, this is not the case. All current and envisioned general purpose supercomputers are based on vector processing. That is, on operations performed conceptually in parallel, on a raise of quantities, "vectors." This restriction allows parallel hardware to greatly accelerate processing speed but also implies that algorithms must be realized either by the author or by automatic software in a vector form. Automatic software is still primitive in this ability so that practically speaking, to achieve the great gains vector processing can afford, algorithms must be recast whenever possible if they are not already vectorized. Fortunately, many common algorithms can be vectorized.

In order to understand this more completely, let us look at an example selected from numerical linear algebra. This is a good choice of area for two reasons:

1. many mathematical models, including the ones discussed above, involve the solution of equations which are reduced to linear in solution process;
2. most algorithms from linear algebra are readily vectorized.

We shall direct our discussion of vectorization at the ETA 2 (CYBER 205) supercomputer. The ETA 2 features memory-to-memory architecture so that vectors are defined as arrays of quantities contiguously stored in main memory. The ETA 2 has vector instructions which perform all the standard arithmetic operations as well as many others which aid in realizing vectors in memory.

Our problem is the solution of the equation Ax=Y for n-tupals x and y and n x n symmetric positive definite matrix A. This is a common application problem which occurs, for example, in finite element methods. The standard solution is to apply to Cholesky decomposition to A in order to write the equation as $LL^tx=y$ for a lower triangular matrix L. Then one only needs to solve LZ=Y, $L^tx=z$.

We assume that only the lower triangle of A is stored in the standard FORTRAN order, i.e., by columns. Thus, each column of A is a vector on the ETA 2. We proceed by observing that if $A=LL^t$, then representing $A = (A_{ij})$ and $L = (L_{ij})$ we have:

$$a_{ij} = \sum_{k=1}^{j-i+1} \ell_{ik}\,\ell_{jk}\ , \ izj$$

It follows that: $\ell_{11} = \sqrt{a_{11}}$ and $\begin{pmatrix} \ell_{21} \\ \vdots \\ \ell_{n1} \end{pmatrix} = \ell_{11}^{-1} \begin{pmatrix} a_{21} \\ \vdots \\ a_n{}^1 \end{pmatrix}$. In general,

if we let (x) $\begin{pmatrix} \ell_{ij}{}^1 \\ \vdots \\ \ell_{nj}{}^1 \end{pmatrix} = \begin{pmatrix} a_{jj} \\ \vdots \\ a_{nj} \end{pmatrix} - \sum_{k=1}^{j-1} \ell_{jk} \begin{pmatrix} \ell_{jk} \\ \vdots \\ \ell_{nk} \end{pmatrix}$ for j>1 and $\ell_{jj} = \sqrt{\ell_{jj}^1}$, then

$\begin{pmatrix} \ell_{jj} \\ \vdots \\ \ell_{nj} \end{pmatrix} = \ell_{jj}^{-1} \begin{pmatrix} \ell_{jj}{}^1 \\ \vdots \\ \ell_{nj}{}^1 \end{pmatrix}$ is the j^m column of L.

Thus, L is computable inductively, the computation consists of mostly vector operations on the ETA 2 and moreover, the basic operation (*) is a sequence of vector operations of the form X-aY for vectors xy and scalar a. On the ETA 2, this is a linked triad and is executed as a single vector instruction.

To continue we must solve Lz=Y. We assume that L is stored in A's space by columns which a closer observation of the above algorithm shows can be done simultaneously with the computation. We have:

$$\begin{aligned} y_1 &= \ell_{11} Z_1 \\ y_2 &= \ell_{21} Z_1 + \ell_{22} Z_2 \\ &\vdots \\ y_n &= \ell_{n1} Z_1 + \ell_{n2} Z_2 + \cdots + \ell_{nn} Z_n \end{aligned}$$

Thus $Z_1 = \ell_{11}{}^{-1} y_1$. If we let $\begin{pmatrix} j_{j+1}{}^{(j)} \\ \vdots \\ y_n{}^{(j)} \end{pmatrix} = \begin{pmatrix} y_{j+1}{}^{(j-1)} \\ \vdots \\ y_n{}^{(j-1)} \end{pmatrix} - Z_j \begin{pmatrix} \ell_{j+1} \cdot j \\ \vdots \\ \ell_{nj} \end{pmatrix}$ for j ≥ 1, then

$Z_j = \ell_{jj}^{-1} y_j^{(j-1)}$. Thus Z can be completed with vector operations with the basic operation being a linked triad.

Finally we must solve $L^T x = Z$. We have:

$$Z_1 = \ell_{11} x_1 + \cdots + \ell_{n1} x_n$$
$$Z_2 = \ell_{22} x_2 + \cdots + \ell_{n2} x_n$$
$$\vdots$$
$$Z_n = \ell_{nn} x_n$$

It follows that:

$$x_n = \ell_{nn}^{-1} Z_n$$
$$x_{n-1} = \ell_{n-1,\,n-1}^{-1} (Z_{n-1} - \ell_{n,\,n-1}\, x_n)$$
$$\vdots$$
$$x_j = \ell_{jj}^{-1} (Z_j - \ell_{nj}\, x_n - \cdots - \ell_{j+1,\,j}\, x_{j+1}) \text{ for } j \geq 1.$$

Now notice that the basic computation to compute X_j is the formation of the dot product of the vectors

$$\begin{pmatrix} \ell_{j+1,j} \\ \vdots \\ \ell_{n,j} \end{pmatrix} \text{ and } \begin{pmatrix} x_{j+1} \\ \vdots \\ x_n \end{pmatrix}$$

This operation is available on the ETA 2 as a single vector instruction. This completes demonstration of a vector algorithm for solving Ax=y. With this algorithm, one can decompose a 500 x 500 matrix in .539 seconds on the ETA 2.

This example indicates that vectorization is a profitable way to achieve some parallelism. However, the next generation of supercomputers, such as the ETA10 and Cray 2 will consist of multiple parallel vector processes. Thus, the future will again require a re-examination of algorithms to discover methods of efficiently using independent communicating processors on a single problem in order to advance the state-of-the-art of mathematical modeling.

References:

1. W. Ballhaus, *Advanced Aerodynamics and Supercomputing,* Proceedings of the Conference "Supercomputers", February 1984.
2. D. Davis, *Supercomputers: A Strategic Imperative,* "High Technology", May 1984.
3. S. McGrogan and G. Tarsy, *Vector Enhancement of a Circuit Simulation Program,* Proceedings International Symposium on Vector Processing Applications, Colorado State University, August 1982.
4. S. McGrogan, *VLSI Circuit Simulation Using a Vector Computer,* Proceedings CYBER 200 Applications Seminar, NASA Conference Publication 2295, October 1983.
5. B. Mossberg, *MAGAV: Math Geophysical Vector Library,* Control Data, 1983.
6. J. Riganati and P. Schneck, *Supercomputing,* "Computer", October 1984.
7. *The Influence of Computational Fluid Dynamics on Experimental Aerospace Facilities,* National Academy of Sciences, Washington D.C., 1983.

CYBERPLUS, a High Performance Parallel Processing System

Wayne A. Ray

Director Parallel Processing
Control Data Corporation
Minneapolis, Minnesota, USA

The CYBERPLUS is a High Performance Parallel Processing System offered by Control Data Corporation and provides a capability for large scale simulations that are not available with other uniprocessor systems or systems containing special purpose array or attached processors.

The CYBERPLUS Parallel Processing System is a high performance parallel computing system which allows the user to increase the scope and size of the simulation by adding additional high performance CYBERPLUS processors. A CYBERPLUS system is an integration of a Cyber 180 Series Computer and one or more CYBERPLUS Processors. A single CYBERPLUS processor is capable of providing up to 700 MIPS (Million Instructions Per Seconds) plus 100 MFLOPS (Million Floating Point Operations Per Second). The CYBERPLUS software allows the simulation user to configure the performance of 1 to 64 CYBERPLUS Processors all operating in parallel on a single simulation.

The CYBERPLUS system contains a unique ring interconnection architecture which allows for very high speed interchange of data and control information between individual CYBERPLUS processors. A ring group contains two 16 bit data rings and a single 64 bit data ring which interconnects 2 to 16 CYBERPLUS Processors. A Control Data Cyber 180 Computer can support up to 18 ring groups, thus a maximum CYBERPLUS system could contain up to 289 independent processors. Each CYBERPLUS ring group connects to the Cyber 180 via a Cyber channel and a high speed direct memory interconnection.

The CYBERPLUS Processor has a 20 nanosecond instruction cycle time and because of its unique instruction processing architecture is capable of initiating up to 15 parallel functions every machine cycle. The functional units are interconnected by a crossbar connection that allows output from one functional unit to be transferred to any or all of the functional units in the machine cycle. With the addition of the floating point unit, the CYBERPLUS processor is able to provide up to 65 MFLOPS in 64 bit floating point bit format or 103 MFLOPS in 32 bit floating point format. This MFLOPS performance is in addition to the CYBERPLUS base processors performance. The floating point unit adds 3 additional functional units (add/subtract, multiply, divide/square root) that operate in parallel with the base processors 15 functional units.

The structure of the CYBERPLUS Processor is built around a very long instruction word. An instruction word contains 240 bits and is read from a high speed bipolar memory every machine cycle. The 240 bit instruction word provides provides the capability of dynamically reconfiguring the 18 parallel functions every machine cycle. Because of its instruction architecture, the machine provides for a very high performance engine plus it allows the user to think of the machine as either a data flow or a data flow like computer system. The architecture of the instruction handling in a CYBERPLUS Processor has many of the properties of both an SIMD and MIMD machine arthitectures.

The memory system of the CYBERPLUS contains four independent bi-polar 16 bit data memories, each memory can read or write 16 bits of data in one machine cycle. A 512K - 64 bit bi-polar memory can deliver three 64 bits of data every machine cycle.

As an example of the parallel instructions, let's assume you wanted to solve an equation where the variables on the right hand side are using the 16 bit data memory and the variable N is stored in the 64 bit memory:

$$N(I) = J(I) + K(I) + L(I) * M(I)$$

for I = 1 to LIMIT. The CYBERPLUS instruction flow would be as follows for the loop initialization and 4 CYBERPLUS instructions.

I	Initialize loop
J, K, L, M	fetch 4 values
J + K, L * M	add, multiply
Add, Increment loop	add
N, Test	store, test loop to 2

Thus, the loop can be executed in 4 CYBERPLUS machine cycles due to the hardware's ability to perform parallel initiation of the independent functional units.

The equivalent performance can be achieved using 32 bit floating point assuming the 32 bit data would be packed in the 64 bit memory word. The loop would be increased to seven cycles to perform the calculations in 64 bit floating point format.

A Matrix Multiply of a 400 x 400 matrix takes 6.25 seconds on a single CYBERPLUS Processor using the 64 bit functional units and less than 3.5 seconds using the 32 bit functional units.

The CYBERPLUS Processor is connected to other processors by a dual 16 bit ring architecture. This dual ring allows for very high speed transfer of data and control information between all processors on the ring. The total ring bandwidth is a function of the number of CYBERPLUS processors since each Processor can be reading and writing a data packet every machine cycle. To move a 16 bit data packet between adjacent processors takes 20 nanoseconds, but because of the ring architecture structure each of the rings can have n packets of data moving around each the ring simultaneously. If the ring consisted of 10 CYBERPLUS Processors, the bandwidth of each of the dual ring would be 8000 MBITS of data.

MINIMUM HARDWARE INTERCONNECTION

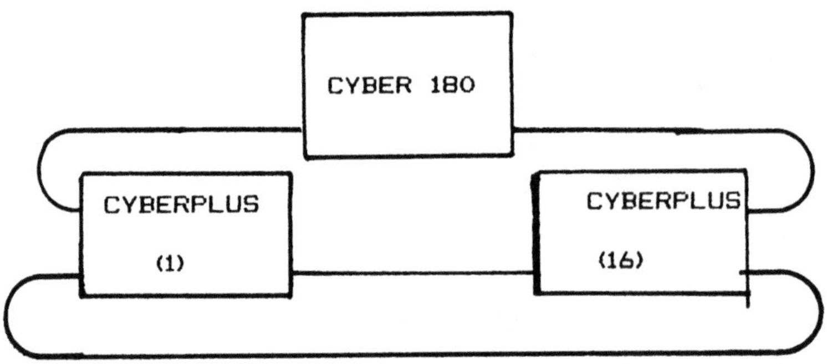

A Cyber 180 Computer can support up to 18 Ring Groups and each Ring Group supports 1 to 16 CYBERPLUS High Performance Processors. The Cyber 180 Computer supports tapes, disks, special devices and communications hardware.

In addition to the small ring architecture, there is a high speed memory interconnection which allows a CYBERPLUS processor to move data and any other processor within its ring group at a sustained speed of one 64 bit word every 20 nanoseconds. The CYBERPLUS Processor can be configured to move data between a CYBERPLUS processor and the Cyber 180 computer at the sustained rate of one 64 bit word every 80 nanoseconds or a transfer rate of 800 MBITS.

This allows the user to have the full range of capability of the standard Cyber 180 operating system, and its associated compiliers, assemblers, and data management. The Cyber Communications capabillity allows the installation to connect the CYBERPLUS System to other computer systems via the networks facilities within the Cyber Operating System.

HARDWARE INTERCONNECT

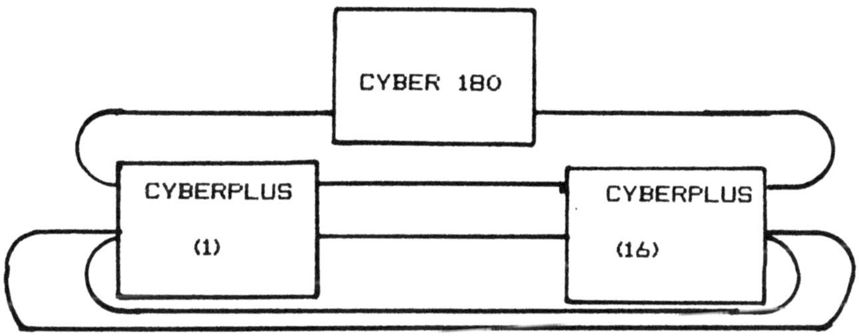

CYBERPLUS SOFTWARE

In order to do simulation applications, the user needs the ability to bring into execution the additional CYBERPLUS processors at the given job step. The simulation user needs to move code to a CYBERPLUS Processor, have the processor execute the code and go about executing other parallel computations in the Cyber 180 or initiating additional CYBERPLUS Processors.

The software architecture allows the Cyber job, the initiator of the CYBERPLUS tasks, to have full control of four CYBERPLUS Ring Groups that have been configured in the total configuration. Thus a user can have control of up to 64 CYBERPLUS Processors. Using a one single CYBERPLUS processor system, the Cyber Processor would obtain data sets, process elements of the data set, then move the data to a computational process that will be executed by the CYBERPLUS Processor. This can be achieved in parallel. The Cyber then has the option to process additional data, to post process previous data or the true case of a parallel processing system allow the application to monitor what is happening in the CYBERPLUS.

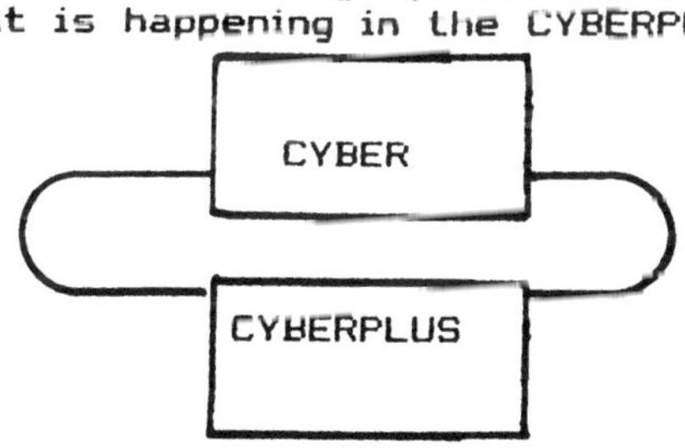

This monitoring operation uses a hardware capability and does not interrupt the CYBERPLUS due to the options built into the CYBERPLUS hardware system. Thus a user could obtain the immediate results from a simulation and use these results for other decision making processes. This decision making process could be the initiation of another CYBERPLUS processor, the initiation of another simulation activity with partial results from a CYBERPLUS or allow the user to abort the process within the CYBERPLUS because we have reached a boundry condition or a condition that longer is valid to the simulation. Thus the user can start to handle parallel simulation activities and not be forced to do the operation in the classic discrete simulation.

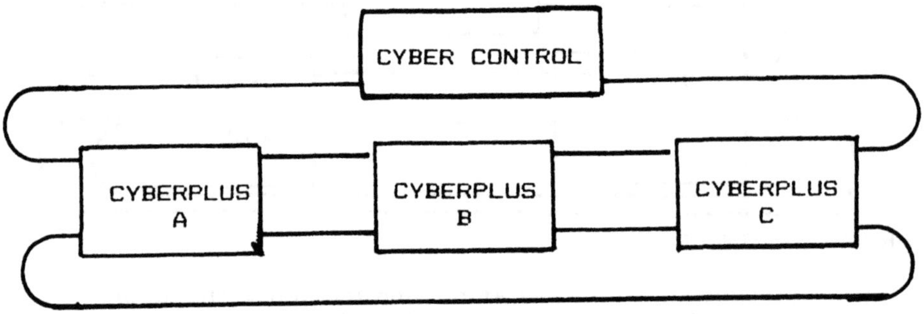

Carrying this one step further, the ring architecture will allow for processors to interchange data such that the user could configure the system to have two simulations going on similtaneously but each of them are transmitting data on a continuous basis between the CYBERPLUS processors in order to simulate a large physical system. Each simulation could then be extended to multiple CYBERPLUS processors to execute each of the initial simulation tasks as the size of the simulation is increased.

One of the problems facing large simulation systems, is the ability to provide the user, the output in a manner that can support user needs. The simulation needs of the future requires a very high sophisticated computer graphic system not just color but color to the nth degree such that the user could initiate the simulation, start to look at critical events, critical operations, boundary conditions, alarms, etc and be able to change the simulation. One or more of the CYBERPLUS processors in the high performance simulation system would have the function of developing the graphic representation. The graphic representation would use the parallel hardware and software architecture and provide the simulation model the capability to be built dynamically. The simulation user would then be able to monitor and understand what it happening in the simulation rather than looking at just the final results.

The CYBERPLUS Software consists of a Cross Fortran ANSI 77 Compiler, A Cross Assembler Debug facilities, a Ring Group Simulator and the user software to initiate and control the full capability of the CYBERPLUS hardware architecture.

Summary

With the CYBERPLUS high performance processor, the ring interconnection architecture and the software to link together computational elements, the CYBERPLUS system offers a unique opportunity for simulation.
Not the simulation of the present physical models, but simulation of the total process. The CYBERPLUS offers simulation capability to do the simulation defined by the total as opposed to how we have been forced to do it with respect to present computer architectures.

Session 2 Simulation – Software – Tools

A Contribution on Calculating Vibrations in Coupled Rigid and Elastic Multibody Systems Applied to a Propulsion System – An Application of the Computer Program MEDYNA

D. KARIUS

MTG Marinetechnik GmbH, D-100 Hamburg, FRG

W. KORTÜM

DFVLR (German Aerospace Research Establishment) - Oberpfaffenhofen, Institute for Flight Systems Dynamics, D-8031 Wessling, FRG

Summary

Taking a ship propulsion system as an example, it will be shown how vibration analysis could be performed by means of the simulation program MEDYNA even in the design phase.
Vibration calculation of a complete propulsion system and thus a system of coupled rigid and elastic bodies is primarily of importance when considering a long, slender ship structure with "weak stern constructions". Such designs are to be found in the field of naval vehicles.
Besides the general eigenvalue analysis the forced vibration due to propeller forces will be considered.

Introduction

For the investigation of the dynamic behaviour of mechanical systems consisting of rigid and elastic bodies coupled by interconnections as springs and dashpots etc. (Fig. 1), the complete equations of motion for these, so-called, *multibody systems* are necessary.
The increasing requirements with respect to the accuracy of mechanical models yield even more complex and detailed multibody systems, comparable with FEM discretisations.
This leads to considerable difficulties even during formulation of the equations of motion.
With increasing number of bodies, the Langrangian equations, usually applied, yield complicated expressions of energy as functions of the generalized coordinates.
Differentiation of kinetic and potential energy causes much effort and can easily lead to errors, which can only be eliminated by extensive calculation checks. For this reason, for-

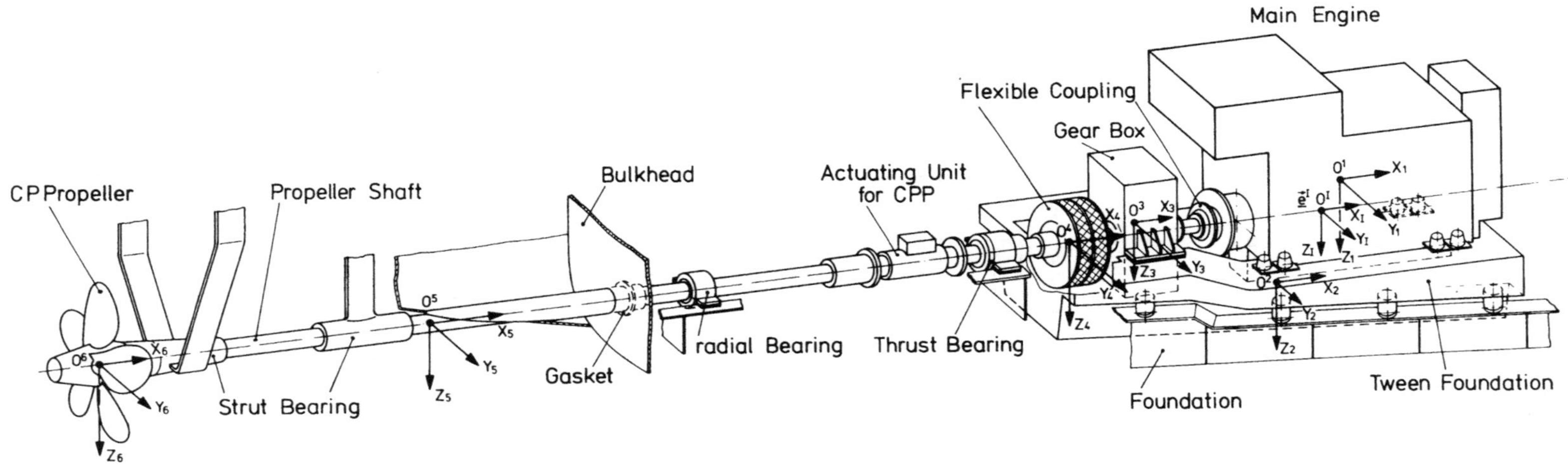

Fig. 1. Propulsion system coupled rigid and elastic bodies

malisms have been developed for automatic calculation of the equations of motion, [1, 2].
For this purpose, d'Alembert's principle is applied to the Newton-Euler equations for the multibody system under consideration. By this procedure, the constraint forces are eliminated and a minimal set of differential equations, the *state-space form*, is obtained.
Usually the equations of motion are obtained *numerically*; however, it has been demonstrated by W. Schiehlen and E. Kreuzer [3], that the equations of motion of large systems can also be determined in symbolic form with the aid of a computer.
For a recent survey on multibody formalisms and programs with emphasis on applications to vehicle dynamics, see [4].
From the equations of motion the desired solutions can then be achieved for linear time-invariant systems by eigenvalue calculation and e.g. application by modal analysis, or for time-variant and nonlinear systems by numerical integration for a fixed set of parameters and initial conditions.

For industrial application the German Aerospace Research Center, DFVLR - Oberpfaffenhofen, Institute for Flight Systems Dynamics, developed an interactive, dialogue oriented program named MEDYNA (MEHRKÖRPERDYNAMIK = multibody dynamics), [5, 6].

Multibody Systems (MBS)

A multibody system is given when the components of the technical system can be replaced by rigid and/or elastic bodies, massless springs and massless dampers, Fig. 1.
The individual rigid bodies, which also can degenerate to mass points, are connected to each other and to their surroundings by various coupling elements as listed in Table 1. This list also includes nonflexible joints e.g. bearings and bars causing constraints.

The global motion of the multibody system, represented by the motion of the reference system $\underline{\vec{e}}^{O}$ ($\underline{\vec{e}}^{O} = [\vec{e}_{\alpha}^{O}]$, $\alpha = 1,2,3$) measured from the inertial system $\underline{\vec{e}}^{I}$, is described by the vector

Table 1. List of linear and nonlinear coupling elements used in MEDYNA

Element	Parameter	Symbol
	linear	
• spring, longitudinal • spring, rotational	k k	
• damper, longitudinal • damper, rotational	d d	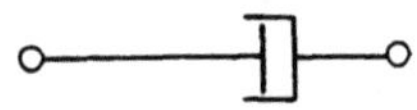
• spring, damper series longitudinal • spring, damper series rotational	k , d k , d	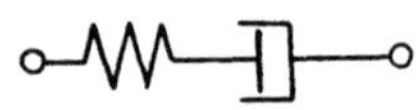
• compact element (6 forces)	$\underline{K}$, $\underline{D}$	
• Beam, bending, torsional beam	EA , EJ GJ	
• dynamic active element $\underline{\zeta}$: state variable $\underline{u}$: control vector $\underline{z}$: measurement vektor	$\underline{H}$, $\underline{A}_{\zeta}$ $\underline{A}_s$, $\underline{A}_{\dot{s}}$ $\underline{B}$, $\underline{C}$	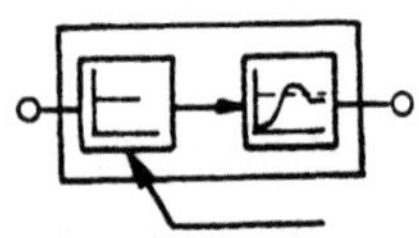
• coupling rod	L	
• general constraints	$\underline{N}_Z$	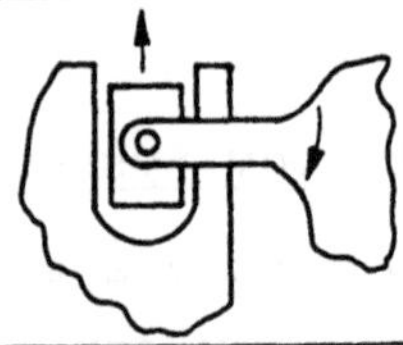
	nonlinear	
• general static/dynamic	USER-Subroutine	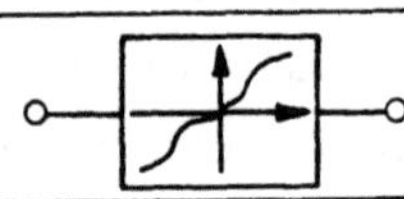

$\vec{\rho}^o$, Fig. 2. If the velocity $\dot{\vec{\rho}}^o$ and the angular velocity $\vec{\omega}^o$ are zero, $\underline{\vec{e}}^I$ and $\underline{\vec{e}}^o$ usually coincide.

In this case the reference point (center of mass) of each individual body is measured from $\underline{\vec{e}}^I$ (otherwise from $\underline{\vec{e}}^o$) by the position vectors $\vec{r}^i_N$ and $\vec{r}^j_N$ to the nominal position of two arbitrary bodies K^i_N and K^j_N of the multibody system, Fig. 2.

In the nominal position N, every body has a nominal system $\underline{\vec{e}}^i_N$ in body-fixed position and is orientated with respect to the reference system $\underline{\vec{e}}^o$. The body-fixed system $\underline{\vec{e}}^i$ in the actual position describes the linearized rigid body motion relative to the nominal position $\underline{\vec{e}}^i_N$ designated by the vector $\vec{r}^i=(\underline{\vec{e}}^i_N)^T\underline{r}^i$ and the Cardan angles $\underline{\gamma}^i = (\gamma^i_1, \gamma^i_2, \gamma^i_3)^T$. For a body-fixed mass-element A^i the position vector $\vec{\eta}^{ia}$ and the orientation matrix $\underline{G}^{ia}$ are given as

$$\vec{\eta}^{ia} = (\underline{\vec{e}}^i)^T\,[\underline{c}^{ia} + \underline{u}^{ia}] \tag{1}$$

$$\underline{G}^{ia} = (\underline{E} + \underline{\tilde{d}}^{ia})\,\underline{G}^{ia}_N . \tag{2}$$

Herein $\underline{c}^{ia}$ contains the vector components of the undeformed body, $\underline{u}^{ia}$ the small elastic deformations, $\underline{G}^{ia}_N$ the nominal orientation, $\underline{E}$ the 3x3 unit matrix and $\underline{\tilde{d}}^{ia}$ a skew symmetric matrix obtained from linearization of the Cardan angle transformation for small elastic deformations.
The small elastic deformations can be represented by the matrix of the mode shapes $\underline{\Phi}^{ia}_{Trans}$, $\underline{\Phi}^{ia}_{Rot}$ as

$$u^{ia} = \underline{\Phi}^{ia}_{Trans}\,(\underline{c}^{ia})\underline{q}^i(t) \tag{3}$$

$$\underline{d}^{ia} = \underline{\Phi}^{ia}_{Rot}(\underline{c}^{ia})\underline{q}^i(t) . \tag{4}$$

The shape functions $(\underline{\Phi}^{ia})$ are obtained from FEM-models or analytical continuum calculations. As a result, the total global position vector for an unconstrained system consisting of n bodies is of the dimension n_p x 1:

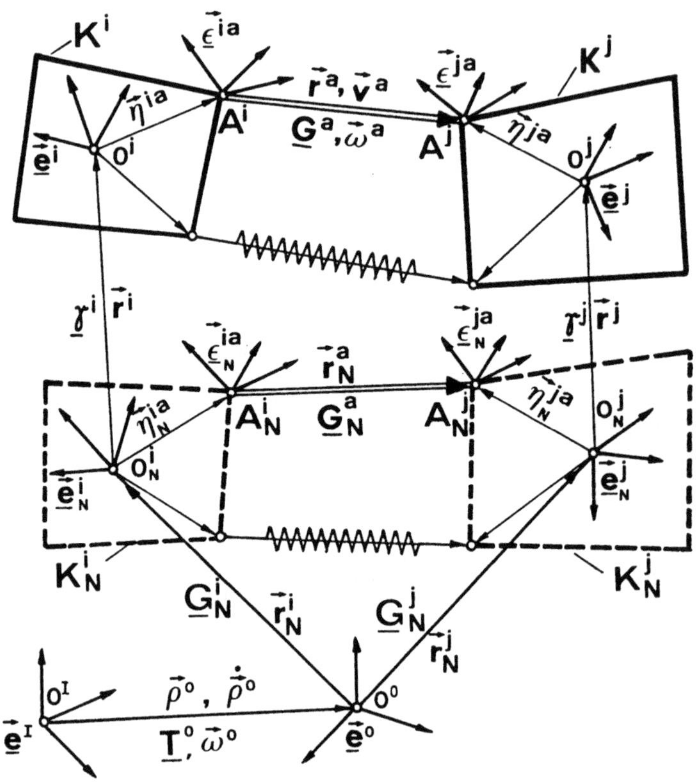

Fig. 2: Reference coordinate systems and unit vectors $\vec{e}^I = (\vec{e}_1^I, \vec{e}_2^I, \vec{e}_3^I)^T$
K_N^i: body in nominal position
K^i : body in actual position
Interconnection index a

$$\underline{p} = \{[(\underline{r}^i)^T, (\underline{\gamma}^i)^T, (\underline{q}^i)^T]^T\}, \; i=1,\dots,n$$

where

$$n_p = \sum_i (6 + n_q^i) \quad i=1,\dots,n.$$

The resulting n_p linearized dynamic equations for $\underline{p}(t)$ in abbreviated form are obtained as

$$\underline{M}\,\ddot{\underline{p}}(t) = \underline{g}(t, \underline{p}, \dot{\underline{p}}, \dots) \tag{5}$$

[6], where $\underline{M}$ is the constant symmetric mass matrix, which contains the mass integrals using the shape functions $\underline{\phi}^{ia}$. The vector $\underline{g}$ contains all generalized forces, which result from elastic deformations, gravitation, Coriolis and centrifugal forces and gyrostatic forces, which are expanded up to linear terms in $\underline{p}$ and $\dot{\underline{p}}$, (Table 2).

Interconnection of multibody systems

The mathematical models for the most important linear and non-linear interconnecting elements used in MEDYNA are listed in Table 1. For linear elements the interaction forces are linear functions of the relative motion between two contiguous bodies. Also for linear holonomic constraints in the multibody system, the $n_p \times n_z$ constraint matrix $\underline{C}_z$ is obtained by MEDYNA, yielding a set of linear relations between the coordinates $\underline{p}$:

$$\underline{C}_z^T \underline{p}(t) = \underline{z}(t). \tag{6}$$

The components of $\underline{z}(t)$ are either zero for locked modes, caused by joints, or they are prescribed time-functions for kinematically excited modes.
If n_z is the number of holonomic constraints, the vector $\underline{y}$ is of the dimension $n_p - n_z$, the number of degrees of freedom of the system. Thus, $\underline{p}$ can be expressed in terms of $\underline{y}$ and $\underline{z}$ as

$$\underline{p}(t) = \underline{J}_y \underline{y}(t) + \underline{J}_z \underline{z}(t), \tag{7}$$

where $\underline{J}_y$, $\underline{J}_z$ are constant Jacobians. The determination of the generalized coordinates $\underline{y}$ and the Jacobian matrices $\underline{J}_y$ and $\underline{J}_z$ is performed by the program, [7].

Calculation

Before starting numerical solutions of the equations of motion, (Table 3), in many applications it is necessary to compute the equilibrium position.
For given nominal static and external forces $\underline{g}_{N1}$ (Table 2), the nominal constraint forces $\underline{f}_{zN}$ and the applied forces $\underline{f}_{nN}$ are computed by

$$\underline{f}_{zN} = \underline{J}_z^T \underline{g}_{N1} \tag{8}$$

and

$$\underline{f}_{nN} = \underline{T} \, \underline{J}_y^T \underline{g}_{N1} ; \tag{9}$$

$\underline{J}_y$ and $\underline{J}_z$ are the Jacobian and $\underline{T}$ the distribution matrices.

Another problem exists, if these forces are known, but the equilibrium conditions are not satisfied.

Table 2 Summary of generalized forces acting on MBS. ($\underline{K}$, $\underline{D}$, $\underline{R}$ (with indices) are constant matrices; $\underline{r}$ denotes additional state components; $\underline{f}_{NL}$ denotes the nonlinear interaction forces and $\underline{f}_{dz}$ the constraint forces.)

generalized forces	nominal	+ linear, time-invariant in $\underline{p}$, $\dot{\underline{p}}$	+ nonlinear or time-varying terms
$\underline{g}$	$= \underline{g}_N$	$- \underline{K}\,\underline{p} - \underline{D}\,\dot{\underline{p}}$	$+ \underline{g}_d(t, \underline{p}, \dot{\underline{p}}, \ldots)$
as a result of			
elastic deformations	$- \underline{K}_q\,\underline{q}_N$	$- \underline{K}_q\,\underline{q} - \underline{D}_q\,\dot{\underline{q}}$	
gravity	$\underline{g}_{gN}$	$- \underline{K}_g\underline{p}$	$+ \underline{g}_g(t, \underline{p})$
relative motion	$\underline{g}_{cN}(\underline{\omega}^\circ{}_N)$	$- \underline{K}_c(\underline{\omega}^\circ{}_N)\underline{p} - \underline{D}_c(\underline{\omega}^\circ{}_N)\dot{\underline{p}}$	$+ \underline{g}_c(t, \underline{p}, \dot{\underline{p}})$
gyrostats	$\underline{g}_{RN}(\underline{\omega}^\circ{}_N)$	$- \underline{K}_R(\underline{\omega}^\circ{}_N)\underline{p} - \underline{D}_R\dot{\underline{p}}$	$+ \underline{g}_R(t, \underline{p}, \dot{\underline{p}})$
applied external forces	$\underline{g}_{eN}(\underline{f}_N)$	$- \underline{K}_e(\underline{f}_N)\underline{p}$	$+ \underline{g}_e(t, \underline{p}, \dot{\underline{p}}, \ldots)$
applied coupling forces	$\underline{g}_{nN}(\underline{f}_N)$	$- \underline{K}_n(\underline{f}_N)\underline{p} - \underline{D}_n\dot{\underline{p}}$	$+ \underline{R}_r\underline{r}$ $+ \underline{R}_{NL}\underline{f}_{NL}(\underline{p}, \dot{\underline{p}}, \underline{f}_{dz}, \ldots)$
constraint forces	$\underline{g}_{zN}(\underline{f}_N)$	$- \underline{K}_z(\underline{f}_N)\underline{p}$	$+ \underline{R}_z\,\underline{f}_{dz}$

In this case the equilibrium position must be found. The determination of the equilibrium requires large body motion; since only linear kinematic relations are considered, a computational procedure is applied in MEDYNA to solve the nonlinear equations iteratively until the given residuals are satisfied. For further analysis, it is useful to rewrite the equations of motion in first-order state-space form. The state-vector then contains the generalized coordinates $\underline{y}$ and its derivatives: $\underline{x}^T = (\underline{y}^T, \dot{\underline{y}}^T)^T$. Herein the constraint forces are eliminated.
This leads to the state-equation

$$\dot{\underline{x}} = \underline{A}\,\underline{x} + \underline{B}\,\underline{u} + \underline{b}(t) + \underline{b}'(\underline{x},\underline{u},t\ldots) \qquad (10)$$

with constant system matrices $\underline{A}$, $\underline{B}$ and the time- and state-variable dependent vectors $\underline{b}(t)$, $\underline{b}'$ (Table 3).
The input vector $\underline{u}$ contains the time-dependent external forces and the time-dependent position variations at the interconnections.

Table 3 **Overall state-equations for the MBS system including subsystems. (The vectors $\underline{f}_{NL}$, $\underline{f}_S$, $\underline{f}_G$ denote coupling forces, S = to subsystems, G = to guideway).**

the derivative of global state vector		constant system-matrix	state-vector	+ time-dependent term	state-dependent term
	$d\underline{x}/dt$	$= \underline{A}$	$\underline{x}$	$+ \underline{b}(t)$	$+ \underline{b}'(\underline{x} \ldots)$
MBS kinematics	$d\underline{y}/dt$	$= \underline{E}$	$\dot{\underline{y}}$		
MBS dynamics	$d\dot{\underline{y}}/dt$	$= \underline{A}_y$	$\underline{y}$	$+ \underline{M}'\underline{g}_{cd}(t)$	$+ \underline{B}_{NL}\underline{f}_{NL}(\underline{x})$
		$+ \underline{A}_{\dot{y}}$	$\dot{\underline{y}}$	$+ \underline{M}'\underline{g}_{Rd}(t)$	$+ \underline{B}_S\underline{f}_S(\underline{x})$
		$+ \underline{A}_{yr}$	$\underline{r}$	$+ \underline{B}_u\underline{u}(t)$	$+ \underline{B}_G\underline{f}_G(\underline{x})$
linear dynamic interconnections $(\underline{f}_d^a, \underline{\xi}^a)$	$d\underline{r}/dt$	$= \underline{A}_r$	$\underline{r}$	$+ \underline{B}_{ru}\underline{u}(t)$	
		$+ \underline{A}_{ry}$	$\underline{y}$		
		$+ \underline{A}_{r\dot{y}}$	$\dot{\underline{y}}$		
nonlinear dynamic interconnections $(\dot{\underline{\xi}}^a)$	$d\underline{x}_r/dt$	$=$			$\underline{b}'_r(\underline{x}, \underline{f}_{NL}, \underline{u})$
nonlinear moving subsystems	$d\underline{x}_S/dt$	$=$		$\underline{b}_S(t)$	$+ \underline{b}'_S(\underline{x}, \underline{f}_S, \underline{f}_{NL}, \underline{u})$
subsystem	$d\underline{x}_G/dt$	$= \underline{A}_G$	$\underline{x}_G$		$+ \underline{B}_{GG}(t)\,\underline{f}_G(\underline{x})$
$\underline{M}' = (\underline{J}_y^T \underline{M}\,\underline{J}_y)^{-1}\,\underline{J}_y^T$, $\underline{A}_y = -\underline{M}'\,\underline{K}\,\underline{J}_y$, $\underline{A}_{\dot{y}} = -\underline{M}'\,\underline{D}\,\underline{J}_y$, $\underline{B}_{NL} = \underline{M}'\,\underline{R}_{NL}$					

Stability of the linearized system can determined from the eigenvalues of the system matrix $\underline{A}$.
Computations in the frequency domain are based on the linear equation

$$\begin{aligned}\dot{\underline{x}}(t) &= \underline{A}\,\underline{x}(t) + \underline{B}\,\underline{u}(t)\\ \underline{y}_M(t) &= \underline{C}\,\underline{x}(t) + \underline{D}\,\underline{u}(t).\end{aligned} \tag{11}$$

The components of $\underline{u}(t)$ can be obtained from the actual disturbance input $\underline{\xi}(t)$ with the aid of time-delays. $\underline{\xi}(t)$ can be given as periodic signals or as power-spectral densities.
The transfer matrix

$$\underline{G}(j\omega) = \underline{C}(j\omega\underline{E}-\underline{A})^{-1}\underline{B}+\underline{D} \tag{12}$$

may be computed for further frequency domain analysis (Bode-plots or output spectral densities). For transient motion and nonlinear problems, the complete equation of motion (10) is numerically integrated by several appropriate Runge-Kutta procedures.

Application to a propulsion system

The essential point of view in this application is to calculate the natural frequencies of the MBS and the response of the coupled structure due to propeller exciting forces. In this case the mechanical model consists of six interconnected bodies, (Fig. 1). Namely: main engine, ground tub, gear box, elastic coupling, elastic propulsion shaft and rigid propeller.
The elastic properties of the shaft are taken into account with the first three mode shapes. The modal mass, stiffness and damping coefficients are obtained by pre-calculations with an FEM-program. In this contribution only the longitudinal and transversal motion and the rotation around the y-axis are considered. The calculated natural frequencies are listed in Table 4. The absolute value of the transfer function is plotted in Fig. 3. Herein a sine-sweep-force is acting on the propeller center of mass.

Table 4. Eigenvalues

Generalized No.	coordinates in	direction	body No.	Name	f_0[Hz]	D[-]
1	DISPLACEMENT	1	2	GROUND TUB	2.67	0.0002
2	DISPLACEMENT	3	2	GROUND TUB	3.5	0.0002
3	ROTATION	2	2	GROUND TUB	5.71	0.0003
4	DISPLACEMENT	1	3	GEAR BOX	121.88	0.0006
5	DISPLACEMENT	3	3	GEAR BOX	120.14	0.0006
6	ROTATION	2	3	GEAR BOX	178.0	0.0002
7	DISPLACEMENT	1	4	RATO-COUPL.	7.23	0.0012
8	DISPLACEMENT	3	4	RATO-COUPL.	6.6	0.0003
9	ROTATION	2	4	RATO-COUPL.	7.32	0.0012
10	DISPLACEMENT	3	3	MAIN ENGINE	2.3	0.0003
11	DISPLACEMENT	3	5	PROP.SHAFT	122.0	0.0006
12	ROTATION	2	5	PROP.SHAFT	178.0	0.0006
13	DISPLACEMENT	1	1	MAIN ENGINE	7.06	0.0008
14	5. MODE SHAPE	-	5	PROP.SHAFT	333.0	0.0102
15	DISPLACEMENT	1	6	PROPELLER	190.0	0.0001
16	DISPLACEMENT	3	6	PROPELLER	320.0	0.0011
17	ROTATION	2	1	MAIN ENGINE	10.7	0.0002

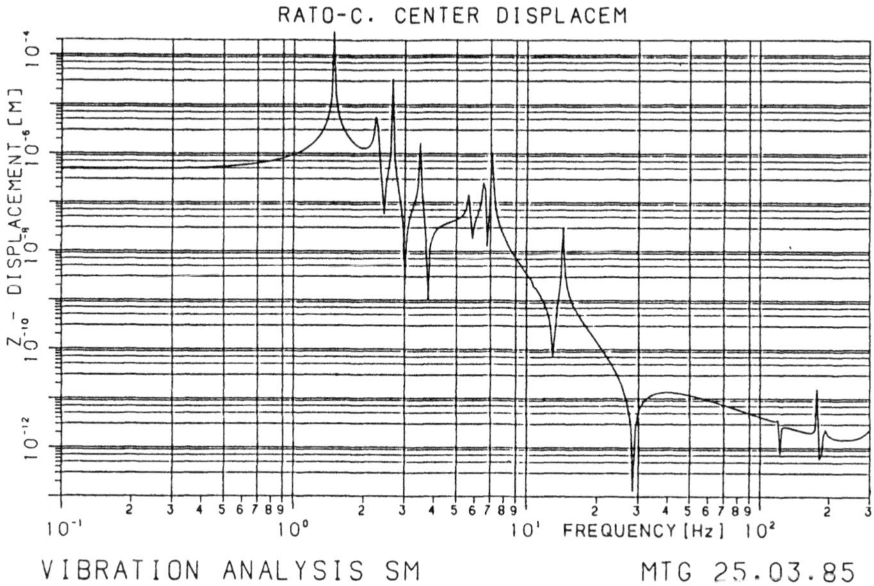
RATO-C. CENTER DISPLACEM
Z - DISPLACEMENT [M]
FREQUENCY [Hz]
VIBRATION ANALYSIS SM
MTG 25.03.85

Fig. 3.

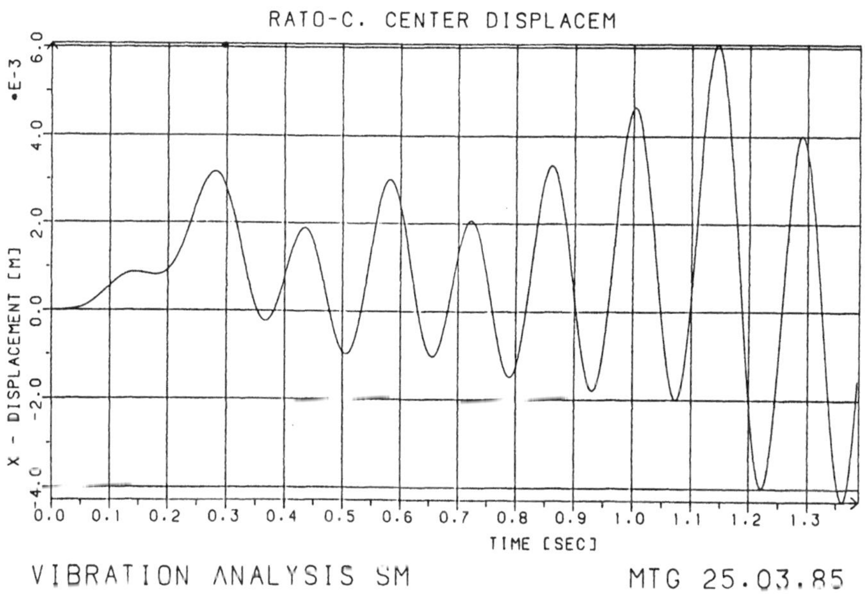
RATO-C. CENTER DISPLACEM
X - DISPLACEMENT [M]
•E-3
TIME [SEC]
VIBRATION ANALYSIS SM
MTG 25.03.85

Fig. 4.

The response belongs to the RATO-coupling center in z-direction. The frequencies of the exciting propeller forces depend primarily on the number of blades and the rotating speed. They are obtained approximately by integration of the two dimensional wake field [9]. Only the forces $\Delta F_x(t) = z \cdot \tilde{F}_x \cos(mz\omega t - \phi_x)$, $\Delta F_z(t) = -z/2\tilde{F}_z \cos(mz\omega t - \phi_z - \pi/2)$ and the moment $\Delta M_y = z/2\ \tilde{M}_y \cos(mz\omega t - \phi_{M_y})$ are considered, wherein z is the number of blades, $\tilde{F}_x$, $\tilde{F}_y$, $\tilde{M}_y$ are the average values, ϕ_i the phase angles and m the m-th harmonic component, [9].

A five blade propeller and only the two first harmonics (m=1,2) are considered. The added mass and damping values belonging to the vibrating propeller, e.g. in longitudinal direction, are approximately given as $m_{prop} = 0.209 \cdot \rho \cdot D^3 \pi F^2/z$ and $d_{prop} = 0.066 \cdot \rho\omega \cdot D^3 \pi F$, [8], wherein ρ is the density of seawater, D the diameter of the propeller and F the nominal expanded area ratio. As an example the vibration due to the propeller forces is plottet versus time in Fig. 4. Herein the displacement of the RATO-coupling center in x-direction is shown.

Conclusion

Due to the growing tendency to support parts of the propulsion system on the hull girders more resiliently,it is becoming more and more important to consider the involved vibrations in the very early phases.
These calculations can be performed very efficiently with the program MEDYNA.

References

[1] Haug, E. J.: Computer Aided Analysis and Optimization of Mechanical System Dynamics, NATO AS, Series, Series F: Computer Systems and Sciences, Vol. 9, Springer Verlag, 1984.

[2] Kortüm, W. (sientif. coord.): Software for Dynamic Analysis and Design of Mechanical Systems, CCG-Lectures V1.08, Oberpfaffenhofen, 02.04. - 06.04.84.

[3] Kreuzer, E.: Symbolische Berechnung der Bewegungsgleichungen von Mehrkörpersystemen. Düsseldorf: VDI-Verlag, Reihe 11, Nr. 32 (1979).

[4] Kortüm, W.; Schiehlen, W.: General Purpose Vehicle System Dynamics Software Based on Multibody Formalisms, to be presented at 9th IAVSD-Symposium on the Dynamics of Vehicles on Roads and on Tracks, Linköping, Sweden, 24. - 28.6.1985.

[5] Wallrapp, O.; Kortüm, W.: MEDYNA - ein Mehrkörperprogramm zur Analyse und Auslegung der Dynamik von spurgeführten Fahrzeugen. VDI-Bericht Nr. 510, 1984, S. 267 - 274.

[6] Kortüm, W.; Wallrapp, O.: General Purpose Software for Vehicle System Dynamics Using Multibody Formalisms, Proceedings First European Cars/Trucks Simulation Symposium, Schliersee, 2. - 4.5.1984.

[7] Führer, C.,; Wallrapp, O.: A Computer-Oriented Method for Reducing Linearized Multibody System Equations by Incorporating Constraints, Journal of Computer Methods in Applied Mechanics and Engineering, Vol. 46, pp. 169 - 175.

[8] Parson, M.G.; Vorus, W.S.: Added mass and damping estimates for vibrating propellers. Soc. of naval arch. and marine engineer. New York (1981), No 16, 273-302.

[9] Schwanecke, H.: Comparison of methods for the calculating of wake induced propeller blade exciting forces. Report of propeller committee 14. ITTC, Ottawa (1975).

Recent Innovations to CSSL-TV

R. N. NILSEN

Simulation Services
Chatsworth, California U.S.A.

Summary

CSSL-IV (Continuous System Simulation Language - version four) provides the most advanced tools available for the modeling and simulation of dynamic systems. Version 4.3, released first quarter this year (1985) has added a complete Linear Algebra Subsystem, several new Graphic Features, over 30 new Simulation Operators, new Integration Algorithms with stiffness detection and automatic algorithm selection, a Linear System Analysis Subsystem plus improved performance and extended availability.

Introduction

CSSL-IV is a user-oriented software system. CSSL-IV is designed to assist engineers and scientists to mathematically model, analyze and evaluate the dynamic behavior of physical phenomena. By providing a set of tools for computer-aided-analysis, CSSL-IV makes it simple for the user to rapidly get his simulation on the computer and to easily conduct experiments, collect data and present that data in useful form with minimal knowledge of the computer system itself.

CSSL-IV is a mature software system having evolved from the original CSSL-I developed in 1968. CSSL-IV contains all the successful features of its predecessors plus numerous enhancements resulting from advances in system theory and technology. CSSL-IV is a proven software system that provides the most advanced tools available for the modeling and simulation of dynamic systems.

The CSSL-IV software system consists of five major subsystems. First, the CSSL-IV model definition language - based upon the SCS standard - enables the user to express his model in user-oriented terms using equation formats and easy to use mathematical operator notation. Next,

CSSL-IV provides a translator which converts the user's model definition into a correct set of FORTRAN-77 programs which are then compiled into computer executable form. A CSSL-IV Software Support Library, containing an extensive set of checked-out utilities for numerical analysis, linear algebra, simulation operators, data collection and graphics, minimizes user code generation. The fourth element is a Run-time Monitor which provides the user interface for controlling the simulation, changing parameters, file management and other interactive facilities. Finally, an operating system interface provides the control and command sequencing required to insulate the user from the details of job sequencing and control.

CSSL-IV has been applied to many diverse areas ranging from robotics to pharmo-kinetics, classical engineering disciplines to socio-economic systems. In the area of marine systems, CSSL-IV has been successfully applied to the areas of: (1) ship dynamics, (2) platform stability and control, (3) fish population dynamics, (4) ocean dynamics, (5) marine propulsion systems, and (6) naval fire control systems. CSSL-IV, as a decision support system, allows the user to readily explore the WHAT IF... nature of various decisions and to explore their consequences before physically implementing them.

This paper will present the new features added to CSSL-IV release 4.3. These new features include (1) a Linear Algebra Subsystem, (2) New Graphics Features, (3) New Simulation Operators, (4) New Integration Methods, (5) New Linear Analysis Support and (6) Improved Performance. New documentation has also been generated.

Linear Algebra Subsystem

CSSL-IV has added a new Linear Algebra Subsystem to assist in implementing models requiring (1) vector algebra, (2) matrix algebra, (3) solutions to sets of linear algebraic equations, and (4) polynomial manipulation. Over 60 new operators and operations have been added to satisfy the increased use of vector and matrix methods in modeling and to make CSSL-IV less dependent on other libraries such as IMSL, EISPAK, LINPAK, etc.

Basic vector operators provide for initialization, declaration and allocation of vectors. The ability to copy, exchange (SWAP), sum components or find max/min component is also included. Arithmetic operations of addition, subtraction, multiplication and division by a scalar constant are easily accomplished. Vector addition and subtraction are also preprogrammed. Vector product operators for both scalar (dot) product and vector (cross) product are implemented. An operator to compute the Euclidean norm (magnitude) of a vector which minimizes overflow and controls round-off error is included. Finally, several tests for comparing vectors for equality, parallelism or perpendicularity have been implemented.

Basic matrix operators provide for initialization, declaration and allocation of matrices. Operators to copy, exchange (swap) and find max/min of row or column are also included. The ability to add or subtract a column or row to other columns or rows is also provided. Multiplicative scaling by a constant is provided by allowing multiplication or division of all elements by a scalar. The product of two matrices as well as the transpose-product are also implemented. Complex matrix operations such as conjugate and transpose conjugate complete this set of basic matrix operators.

Combined matrix-vector operators allow for arithmetically combining vectors and matrices. Special operators compute the inverse, determinant, eigen values and characteristic equation of a matrix. An operator to compute the LU decomposition in preparation for the solution of matrix algebraic systems (Ax=b) is also provided.

Finally, a set of operators to perform polynomial manipulation have been added. These operators find the roots of a polynomial, evaulate a polynomial, and create a polynomial from its roots. Polynomial arithmetic is also provided to add or multiply pairs of polynomials.

New Graphics Features

Several new graphics features as well as extensions to existing features have been released. CSSL-IV supports a wide range of graphics devices, graphics packages and graphic data bases.

CSSL-IV introduces run-time graphics through the use of the RGRAPH directive. When RGRAPH is entered followed by a START directive, the graphic output is generated "during the execution" of the experiment. This is in contrast to the GRAPH directive which generates graphic output after the experiment and from a data base generated during the experiment. The dynamic range of the x-and y-axes must be specified prior to using RGRAPH in order to scale the graphic output. Subdirectives SCALEX and SCALEY are used for this purpose. Another feature of RGRAPH is that it is interruptable. The CSSL-IV monitor supports the "break" or interrupt sequence when issued during the experiment. The experiment is interrupted and control is transferred to the CSSL-IV monitor where parameters can be modified and the experiment restarted or continued.

Two new subdirectives have been added to the PREPARE directive - namely SAVE and RESTORE. PREPARE SAVE empties the graphics buffers to the disk and saves the plot data and name tables so that at some later time (tomorrow, next week or next month) the user can restore these files and generate graphics without re-running the experiment. PREPARE RESTORE is used to initiate the file manager and restore the graphics data files. Also added to the PREPARE directive (as well as to OUTPUT and RANGE) is the ability to concatinate names. For example, PREPARE T, X, Y will direct CSSL-IV to collect data on T, X & Y during an experiment. However, if a name (Z for example) is to be added to ths list, one has only to enter an additional directive -PREPARE Z- to add Z to the list. To clear the list, enter PREPARE with no arguments.

CSSL-IV has added the GKS (Graphic Kernal System) to the list of graphic libraries supported by the code. CALCOMP, TEKTRONIX PLOT/10/TCS, VERSATEK and DISPLA continue to be supported by CSSL-IV.

Finally, several new graphics subdirectives have been added. Choice of axis-type can now be made using one of the subdirectives LINEAR (default), SEMILOG, or LOGLOG. Choice of grid style can be selected using one of the subdirectives GRID (default), OPENGRID or NOGRID. Lastly, the physical size (where applicable) and axis scale factors can be explicitly specified using the subdirectives SIZEX, SIZEY, or SCALEX, SCALEY.

New Simulation Operators

CSSL-IV introduces over 30 new simulation operators. New functions have been implemented in the areas of (1) Linear System Operators, (2) Non-Linear Operators, (3) Sampled-data Systems, and (4) Analysis Support Operators. Simulation Services continues to respond to the needs of the user.

The number of linear system operators has been extended to cover a more extensive list of Laplace Transforms. CSSL-IV allows the user to define certain linear operators in the s-domain and CSSL-IV automatically translates these operators into an equivalent set of differential equations. New operators include a first order lag - LAG1, a single pole with a single zero - PLZERO -, a double pole - DBLPOL -, a pole plus integrator - SPOL1 -, a band-limited differentiator - DERIVS -, a filter - FLTR12 -, and a double limiting integrator - DBLINT -. A generalized transfer function in pole-zero form - TRANR -, is also provided.

New nonlinear operators have been added to support time domain modeling in several areas. For mechanical systems, operators for static and coulomb friction along with backlash, bang-bang, and relay operators have been added. For vehicle dynamics, the ROTATE and quaternion - QUATRN - operators were developed. For hydraulics, the ABSQF simulated flow as $q|q|$. Finally, a PID (proportional-integral-differential) controller has been added to CSSL-IV.

In the area of sampled-data systems, CSSL-IV has added new sampled-data filters DF1 and DF2 based upon finite difference algorithms. A delay - ZDELAY - is added as a model of Z^{-1}. Algorithm zero was also added to update the derivative block and independent variable for simulations with no integration operators.

Finally, several operators in support of simulation analysis were added. Operators to obtain bode, root locus and gain and phase plots from a given transfer funtion were added. A discrete fourier transform (DFT) and its inverse (IDFT) were added as part of the FFT subsystem. Operators to solve for the zeros of non-linear functions and for multiparameter optimization have also been added to CSSL-IV.

New Integration Methods

CSSL-IV continues to offer the most advanced numerical methods commercially available. Recently added to the list of CSSL-IV numerical methods is the variable-step Runge-Kutta-Fehlberg integration method with stiffness detection. Also added was an extension to the stiff integration subsystem which automatically detects stiffness and automatically switches algorithms depending on the local stiffness of the problem.

The Runge-Kutta-Fehlberg algorithm (algorithm 40) provides access to a fourth-order, self-starting, single-step method with automatic step-size control. The automatic step-size control is based upon error estimates formed by knowing both a 4th order and 5th order per step solution. The key is that Fehlberg found a set of coefficients such that only one additional derivative evaluation is required to get the error estimate. This algorithm also has a heuristic which detects when the problem being solved is "stiff" and issues a warning message suggesting the user switch to the stiff integration method.

Four new variations on the Hindmarsh stiff integration method have been added. These new methods make use of a stiffness detection method developed by Petzel at Sandia Laboratories. At each integration step, a stiff or non-stiff integration method is selected based upon the result of a stiffness detection algorithm. This collection of new methods is therefore variable-method, variable-order and variable-step. The variations depend on whether the Jacobian is user supplied or numerically generated and whether the Jacobian is banded (sparse) or dense.

New Linear Analysis Support

With increased access to digital computers, many of the theoretical aspects of linear systems analysis are now being implemented for large practical systems. CSSL-IV has added several new tools to support simulations using linear system theory. By linear system, we mean a simulation defined by the vector equations

$$x(t) = A\,x(t) + Bu(t)$$

$$y(t) = C\,x(G) + Du(t)$$

where A, B, C, D are matrices and x, x, u, y are vectors. CSSL-IV has added tools to (1) linearize a system about a set point, (2) find the steady state solution, (3) handle the algebraic solution to the above equations and (4) support matrix transfer-function manipulation.

CSSL-IV now has the ability to linearize a system about a set-point. Linearization about a set-point is useful because much can be derived about the system behavior and in particular about the stability of the system under conditions of small-signal perterbations (i.e., linear approximation at an operating point). CSSL-IV provides access to the tools necessary to numerically generate the A-matrix and generate the linearized equations which represent the small signal behavior of the system at a given operating point.

CSSL-IV also provides a steady-state-solver. Many times, the user is more interested in the steady state conditions than in the transient response prior to steady state. CSSL-IV provides the software tools to easily achieve the steady state conditions (i.e., when x=0) without having to recursively solve the differential equations. Algebraic techniques are used to find that state x such that x = f(x,t) = 0.

CSSL-IV has added a set of software tools to support the analytic solution of the above linear description. These tools include an eigenvalue generators, a state transition matrix generator, and the ability to get the characteristic equation, the resolvent matrix, the determinant, the inverse plus other matrix analysis tools.

Finally, CSSL-IV has added the capability to handle and support matrix transfer functions in the frequency domain. The ability to take the A-matrix and generate the open-loop and/or closed-loop transfer function is provided. Support for matrix transfer function analysis is also included. These features are coupled with existing Bode, Nyquist and root-locus methods to generate graphical indications of system performance.

Improved Performance

The developers of CSSL-IV continue to improve the performance of the CSSL simulation software system. CSSL-IV portability has been expanded to over 24 different operating systems. Execution time has been minimized through the use of efficient algorithms with emphasis on numerical accuracy, accuracy control and user warnings of impending problems.

CSSL-IV has provided built-in multiple precision in all numerical methods in order to maintain numerical portability across the spectrum of computers with word-lengths varying from 16-bits to 64-bits in size. Where possible, automatic error detection and control is implemented and tests warn the user of loss of accuracy due to round-off problems. CSSL-IV continues to develop what amounts to an expert system for numerical analysis.

Interactive CSSL-IV also supports the interrupt or "break" function. This means that during any simulation, the user can interrupt the experiment, return to the CSSL-IV monitor, modify a parameter and restart the experiment. This feature proves very useful in large simulations or long simulations because it allows the experiment to be terminated before the end of the experiment with control returned to CSSL-IV and not the operating system.

Conclusions

CSSL-IV continues to provide the most advanced software tools for the simulation of dynamic systems. In this paper, the CSSL-IV simulation software system has been described and the new features added to release 4.3 have been presented. New operators to handle Linear Algebra (vector, matrix and polynomial systems) along with many new simulation operators have been added to CSSL-IV. New graphics features including dynamic graph generation were also added. Several new integration methods provide CSSL-IV with the most advanced numerical analysis subsystem currently available. The Linear Analysis Support System provides for steady-state estimation, linearization at a set point and matrix transfer function capabilities. Finally, the performance of CSSL-IV continues to be improved in the areas of accuracy, compute-time, and user diagnostics.

CSSL-IV is a proven product with over 16 years of evolutionary development. It is fully supported world-wide on over 24 different operating systems plus being available on CDC CYBERNET. CSSL-IV has been successfully applied to many sophisticated and diverse problems including the many aspects of marine systems.

References

Nilsen, R. N., CSSL-IV User's Guide and Reference Manual, 1984.

Strauss, J.C., et al., "The SCi Continuous System Simulation Language (CSSL)", SIMULATION, Vol. 9, N. 5, Dec. 1967, p. 281-292.

The Role of Computer Graphics in Maritime Simulation

D. RIDDER

Control Data GmbH, München, Germany

Summary

Computer graphics constitutes a very powerful and direct interface between a simulation system and its operator. The constraints imposed on a computer graphics system by the special demands of maritime simulation are not very severe. Motions are mostly slow, however, a multitude of display devices is necessary. Depending on the amount of realism wanted some special features for scene display can be realized. Special systems for radar and control-system simulation need only line graphics and standard techniques.

The Importance of Computer Graphics

Any digital simulation system consists of five parts (see Fig.1.). Mathematical models are necessary to describe properly the behaviour of the physical system to be simulated. Sophisticated mathematical methods and algorithms are used to perform the operations prescribed by the model. An appropriately chosen hardware and software system is the basis for the implementation of these algorithms under the conditions of proper timing and accuracy. Next, a very important role is played by the interface to the real world, i.e. to those parts of the simulation system that are identical to the real system. The latter may either be original parts of equipment or the operator/trainee himself. The operator is considered to be the final 'piece' of the whole system. His means of interaction with the system are manifold, but in modern systems the visual interaction prevails. There may be scenes to observe, targets to gauge or meters to read. The necessary input to the operator is mostly generated by computer graphics displays, either as realistic scene displays, enhanced windows, schematic pictures or some mixture of alphanumeric and graphic representation. In the

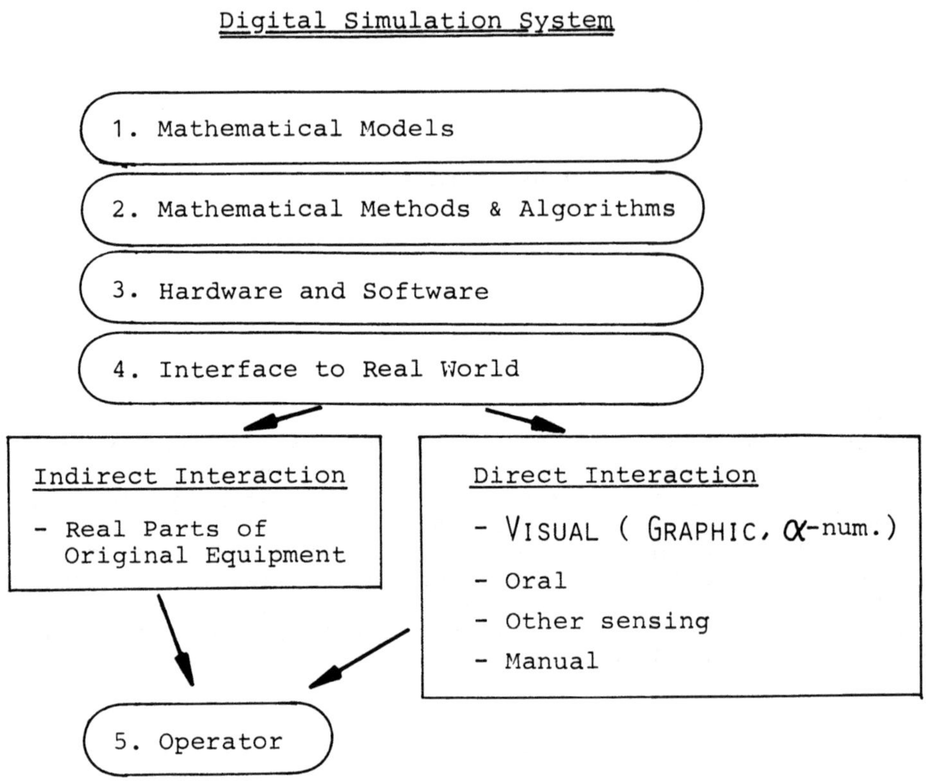

Fig.1. Components of a digital simulation system and the role of visual interaction between operator and system.

course of the evolutionary history, the eye has evolved from a part of the brain. Hence, it constitutes the most direct input channel to our brain. This fact and the observation, that graphical perceptions are interpreted and transformed very effectively in a part of our brain which is specialized in emotional and unconscious thinking [1], manifest the great importance of the graphic interface in simulation systems. Among the other ways of interaction between man and the system, the most important one is mechanical handling of some devices. Finally, also oral and other sensing, e.g. equilibrium sensing, are sometimes useful.

Typical Features of Maritime Simulation

As the whole field of maritime simulation not only consists of training simulation, but also of simulation for details of water flow, ship motion, ship performance and general traffic conditions, the spectrum of the necessary computer graphics features is extremely broad. For this reason I will focus more on those features that are typical for the maritime training case and, hence, will omit those uses of graphics which are of more general nature.

When looking for typical features of maritime simulation, especially in real-time training, one has to consider first the constraints imposed by velocities and dimensions. The velocities of most ships are relatively slow. There are not the heavy constraints as e.g. in flight simulation. So, in normal driving mode representing relative motions of large ships on the high seas or motions relative to some coastline, the timing problems for the display are of minor importance. It may, however, be desirable to include some fast motion effects in order to study ship maneuvering on a reduced time scale. Another feature requiring quick image generation would be the simulation of the motion of the sea especially for smaller vessels. Also for speed-boat simulation, the proper timing will result in more severe requirements on the display hardware and software.

Linear dimensions in the maritime case are generally quite large. Due to high inertia, the ships have to keep at large spacings. This permits a relatively simple kind of representation of other vessels. The appearance of distant objects should be simulated properly by blending them in with continuously increasing detail. Sometimes one may want to simulate the use of telescopes. This can be achieved by zooming transformations which have to be incorporated in the display-software. In this case a geometric representation again with increasing detail is necessary. Another consequence of the large dimensions is that apart from maneuvres in a harbour the 2D-projections to the screen may be simplified. During harbour maneuvres one will need a detailed visualization of the buildings and the machinery on the piers. This may be realized by digitized photographs stored on videodiscs for fast access.

The dimension of distance also plays an important role on the ship itself. The bridge is a quite large room with many windows which need to be simulated. A large number of displays will be necessary to operate efficiently in parallel. This has direct implications on the organization of display-file(s) and the picture database. In order to achieve the proper illusion at manageable cost, it is often necessary to produce part of the picture, e.g. parts of the own ship, by projecting slides onto a screen instead of using expensive computer displays.

Due to the large mass, ship motions have a high inertia and require a long-range preplanning. This demands driving by instruments, especially by radar. So, a proper computer graphic representation of modern radar equipment is an important part of most maritime simulators. The task to simulate modern radar systems is quite challenging, as these systems themselves already involve many computerized functions. The necessary graphics tool, however, may require only vector-screens and line-drawing software. The picture rendering will be a standard graphics problem, but the special feature will be the connection with the picture database used for scene display.

Finally, there are many typical details to be represented in a special maritime graphics system. The most typical is the effect of the motion of the sea on a ship mentioned above. Waves in the vicinity of the ship may need a realistic representation and effective algorithms for their deformation and motion. Weather conditions are still of large influence on shipping. So, one needs proper presentations of fog, stormy weather and even of the sea covered with ice. A more refined software could take into account the varying illumination by the sun, effects of cloudy sky, and also the scene at night with the proper zodiacal figures. Aside from these realistic features necessary for scene display, schematic displays of engine operation etc. should also be realized using standard graphics components.

State of Art in Computer Graphics

For our purpose the field of computer graphics may be subdivided into schematic representations and realistic pictures (see Fig.2.). As indicated above, schematic pictures will be used for displays of simulated radar scenes and systems operations. They require simple features like line-drawing, polygon-filling and alphanumeric labeling. A selected range of colors is mostly sufficient. Properly chosen colored background for different messages can improve the perception by the operator. The main task in designing schematic graphics lies in the use of proper symbols according to the sentences of semiotics and the observation of certain graphics-design rules [2]. Especially the selection of colors has a large impact on the ergonomics of such displays [3].

The goal of realistic computer-generated pictures is to produce the illusion of the real scene [4]. There are basically five steps to achieve optimum realism. Any object has to be described by a mathematical formulation and entered into the computer. For display on a screen, the object has to be projected from its 3D model space to the 2D-plane of the screen. Third, the visible lines/surfaces have to be selected for display. Then, an illumination model is applied to generate the color and shade for each surface. Finally, the appropriate intensities of red, green and blue have to be determined for display on the monitor.

In addition to the usual analytical geometries the entities in our case will require mathematical representations of free-form surfaces, e.g. for ship hulls, and - as the opposite extreme - algorithms to generate geometries with some kind of randomness like waves, clouds, etc.. The former may be treated as Bézier- or B-spline-surfaces [5], whereas for the latter the term fractals is used [6]. A Bézier-surface is represented by a mesh of control points which is smoothed by a deliberately slowly approximating function of two parameters. This procedure assures smoothness of the surface which has only the corners in common with its defining point mesh. When a certain geometrical representation is considered, one should also take into consideration the operations of hidden-surface elimination or shading which have to be performed. It is often more helpful for a fast display to use a representation in terms of plane facets or quadric surfaces which are easier to transform than the parametrically defined smooth surfaces.

Fractal geometries are defined by algorithms which allow for increasing detail and incorporate variations of shape according to random numbers. Such geometries are useful for the generation of many objects in nature, such as water-waves and clouds. Althaough the algorithms yield compact geometric descriptions, they tend to be time-consuming.

Many algorithms exist for the problem of visible surface detection [7]. In general they have to perform a kind of geometrical sorting according to the depth of each picture-element (pixel). The simplest example is the depth-buffer algorithm which, for each pixel, compares all surfaces which may contribute to its intensity. More efficient algorithms take advantage of the fact that most neighbouring pixels have similar properties. This coherence can be used to avoid repeated calculations for these pixels. The basic algorithms of this form are the priority algorithm, the subdivision algorithm and scan-line algorithms. Several improvements of these exist and a comparison of the ten most frequent algorithms is found in the literature [8].

The proper impression of the form of a surface is achieved by appropriate shading. The shading is determined by the distribution of light sources, by surface properties and by the viewing position. Considering that a large scene is displayed and the position of the operator is constrained to the simulated bridge, a constant viewing position may be assumed. Realistic light sources for scene illumination are more difficult, because not only the direct sunlight, but also the averaged sum over all multiple reflected light rays as a kind of background has to be taken into account. Very intricate models (e.g. Torrance-Sparrow-model) have been developed in order to allow for the representation of surface roughness in terms of reflecting

Computer Graphics

Schematic Pictures

- Line Drawing
- Polygon Filling
- Labelling

- Semiotics
- Design-Rules

Realistic Pictures

- Geometric Models
  Analytic
  Free Form
  Fractal
- Perspective Transformations
  Projection
  Perspective Division
  Clipping
- Visible Surface Algorithms
  Depth Buffer Algorithm
  Priority Algorithm
  Subdivision Algorithm
  Scan-line Algorithm
- Shading
  Ambient, Diffuse, Specular Reflected Light
  Phong, Torrance-Sparrow
- Color Representation
  Color Spaces
  Transformations, Interpola-
- Advanced Features
  Anti-aliasing
  Shadows
  Surface texture

Fig.2. Outline of computer graphics features important for maritime simulation.

microfacets. Elaborate shading will be necessary in maritime simulation only for near-by views of other ships during harbour maneuvering and possibly for the surface of the sea.

The final step in realistic image synthesis consists of the determination of the relative intensities of the three primary colors red, green and blue in order to match the desired color impression [3]. This task is rather complicated when taken seriously as the color impression depends not only on the displayed picture but is also influenced by surrounding objects and by the average room illumination.

Some advanced features are mentioned in Fig.2, but only the anti-aliasing will be of importance in our case. Anti-aliasing techniques eliminate stair-case effects inherent to raster display devices by smoothing the surface edges.

As illusion is the ultimate goal of any simulation, it is very important not to destroy it by neglecting certain human factors. The graphic representation has to be comfortable enough to be accepted for reality and must not be disturbed by inappropriate equipment or software. Especially flickering of displays causing eye-fatigue must be avoided, the resolution must be sufficient, the colors have to be in the proper range, and they must not be overemphasized. In the case of schematics, the outfit has to be appealing.

Concept of a Computer Graphics System for Maritime Simulation

A comprehensive graphics package for maritime simulation should involve three parts: scene display to simulate the windows on the bridge, radar simulation, and simulation of ship control systems. It has to include realistic graphics as well as schematics. Scene-display and radar simulation have to work in parallel and must have access to a common picture database. Multiple screens are necessary in order to represent the many windows of a bridge. The software- and data-structure should be adapted to the display system in order to optimize system response. Many geometric transformations will affect all displays in common and may be performed in parallel. Actions like clipping to view borders of a single screen have in this case the meaning of a shift to the next display screen. A specialized display-software may perform such operations very effectively and allow for high performance concerning real-time behaviour and detail.

The rendering system should include the necessary amount of realistic features such as e.g. shading and surface representation. Special effects may be needed to generate different weather conditions and even a representation of the night sky. Existing graphics packages fulfill many of the requirements and may be used in connection with the proper software interfaces to the simulation software for the reproduction of the desired real-time animated pictures.

References

1. Sinz, R.: Gehirn und Gedächtnis. Stuttgart, New York: Gustav Fischer Verlag 1981.

2. Matthaei, J.M.: Grundfragen des Grafik-Design. München: Verlag Moos und Partner 1985.

3. Greenberg, D.; Marcus, A.; Schmidt, A.H.; Gorter, V.: The Computer Image: Application of Computer Graphics. Reading/ Mass.: Addison-Wesley 1982.

4. Hall, R.A.; Greenberg, D.P.: A Testbed for Realistic Image Synthesis. Comp. Graphics and Appl. 3 (1983) 10-20.

5. Rogers, D.F.; Adams, J.A.: Mathematical Elements for Computer Graphics. New York: McGraw-Hill 1976.

6. Mandelbrot, B.B.: The Fractal Geometry of Nature. New York: Freeman and Company 1982.

7. Pavlidis, T.: Algorithms for Graphics and Image Processing. Berlin, Heidelberg, New York: Springer-Verlag 1982.

8. Sutherland, I; Sproull, R.; Schumacker, R.: A Characterization of Ten Hidden-Surface Algorithms. ACM Computing Surveys 6 (1974) 1-55.

Session 3 Industrial Computer System – The Chrysler Story

Simulation Tools for Chrysler Product Development

R. A. Brauburger

Chief Engineer
Technical Computer and Instrumentation Center
Chrysler Corporation, Detroit, Michigan

SUMMARY

Chrysler Corporation's product design and development efforts rely heavily on an extensive network of computers for computer-aided design, engineering, manufacturing, and testing (CAD/CAE/CAM/CAT). Since virtually all proposed vehicle goemetry is stored on the system, it can be used for engineering applications such as structural analysis, crash simulation, and aerodynamic computations. This paper describes the computer environment at Chrysler and the techniques being developed to support crash and aerodynamic simulation on Chrysler's CYBER 205 computer.

TECHNICAL COMPUTING AT CHRYSLER

For more than a quarter of a century, Chrysler has acquired, expanded, and applied computer systems to engineering and scientific computations. Chrysler became one of the pioneers of the Computer Age when it made a large investment in a first-generation mainframe for engineering computing. The system was very advanced for its time but would be considered puny today, when many children's computers have more power. With numerous upgrades and expansions over the years, the system has grown to provide hundreds of millions of instructions per second and nearly 200 gigabytes of rotating mass storage.

The system now contains 27 Control Data Corporation CYBERs, ranging from CYBER 825s to a CYBER 205, organized into several clusters that directly share rotating mass storage. All the CYBERs are linked together by a loosely connected network (LCN) that supports high-speed data transfer. The network is a cohesive system with distributed, shared data bases that allows users quick and easy access to any network data. A CALMA network, an IBM 4341, a Computervision CDS 4000, and two remote CYBER 825s complete the network.

More than 1,100 terminals are attached to the network. About half are high-resolution graphics terminals, such as the Tektronix 41XX, communicating at a 19,200 baud rate. Such terminals provide rapid response and combine with the massive computer network to support a wide range of applications. To minimize response time and to maximize user productivity, interactive applications are moving selected functions, such as graphic control, from the host computer to intelligent terminals.

THE COMMON DATA BASE

The graphics terminal is rapidly replacing the drafting board as a designing and drafting tool at Chrysler. The vast majority of designs in all departments is now assisted or accomplished by computer graphics. A common data base of geometric information such as points, lines, point sets, surfaces, solids, drawings, and many other data types is available to all applications to store and share data. For example, a wire frame or a solid model created by a designer may be used by an engineer for finite element modeling and analysis.

FINITE ELEMENT MODELING

Chrysler began computer-supported structural analysis more than 20 years ago and has employed NASTRAN as its primary analysis code since 1972. In 1975, to facilitate model building and the interpretation of analytical results, Chrysler began the development of GCSNAST (1), its interactive graphics system for the pre- and post-processing of finite element modeling and analysis.

Through close cooperation and continual interaction between users and software developers, GCSNAST has evolved into a friendly, powerful, and efficient tool. GCSNAST has a complete set of finite element modeling functions in 3-D graphics for model creation, manipulation, editing, and verification. It offers complex surface and solid element generation capabilities with automatic mesh creation. GSCNAST handles requests for analysis with a single command, and it evaluates results through a direct link to MOVIE·BYU (2), producing a wide range of pictures such as geometry deformation plots, animation of mode shapes or deformation, color contour plots of stress or displacement, and realistically shaded surface model images. Access to the common data base provides GCSNAST users with geometry created by the design and drafting applications.

CRASH SIMULATION

For obvious reasons, crash-worthiness is a critical concern for automotive designers. Until recently, Chrysler depended on destructive testing and sophisticated impact analysis techniques, including on-board data acquisition systems, high-speed film recording, and computer analysis, to test and optimize vehicle crash characteristics. This approach has some drawbacks. New parts in a proposed design are relatively expensive to create--in fact, a prototype crash vehicle may cost up to one hundred times as much as a production vehicle. New parts and vehicles also require a significant time expenditure to create.

For Chrysler, recent developments in computer hardware and software provide simulation tools that promise to reduce crash testing of vehicles. Grumman Aerospace Corporation has developed DYCAST (3), a crash simulation program, that uses the finite element method to describe the structure while allowing experimental data to be incorporated into the model where needed. Since crash response is highly nonlinear, several hundred iterations, each equivalent to a single linear analysis, are required for a single crash. This analysis requires a tremendous number of computations, necessitating the use of today's most powerful computers. Chrysler uses the CYBER 205 for crash simulation with DYCAST because of the CYBER's inherent speed in scalar and vector mode as well as its large amount of real memory. To date, the solution phase has been vectorized, but element formulation is done in the scalar mode.

Chrysler builds models for crash simulation interactively with GCSNAST on any CYBER except the CYBER 205, which is used exclusively as a batch machine. When DYCAST analysis is requested, GCSNAST generates and sends the appropriate batch job to on the CYBER 205. After analysis, results are routed back to the originating machine. GCSNAST interprets and presents the DYCAST output in a variety of ways, including animation of selected frames of the crash on the high-performance 3-D Evans and Sutherland systems. These systems have local processing capability to allow dynamic rotation, depth cuing, and Z-axis clipping. The animation capability has proven to be essential in eliminating modeling errors as well as visually presenting final results.

Chrysler has begun using DYCAST for head-on barrier crashes. Typical modeling techniques and results are presented below for the analysis of Chrysler's mini-van that was in production before DYCAST was available. The purpose of the mini-van analysis was to develop confidence in DYCAST and in our ability to use it to predict crash behavior. Although a vehicle is not truly symmetric about its mid-plane, symmetry is assumed for this type of test to save computational time. The model consists of 325 grid points, 124 beam elements, 324 triangular membrane elements, 68 nonlinear spring elements, and numerous concentrated masses to bring the total vehicle weight up to curb weight. The resulting model has 827 degrees of freedom with a semi-bandwidth of 536.

Beam elements are used to model the major structural components such as rails, sills, and roof supports. Although the use of beam elements makes it more difficult to represent localized behavior such as section collapse at "hinge" points, the advantage of fewer degrees of freedom and the resulting reduction in computation time over an equivalent shell element model is thought to outweigh this difficulty. DYCAST has eleven standard beam sections ranging from solid circular to hat shapes. In addition, it has a general beam section that is essential for automotive applications since many typical automotive sections are very irregular in shape. Most of the beams in vehicle models use this type of section.

The nonlinear spring element allows the user to define a stiffness connection between points. This stiffness is defined by an arbitrary force deflection curve that may have been determined from either component testing or supplementary finite element analysis. In a vehicle model, these elements are used to represent such things as beam section collapse (rotary springs), tire/wheel crush characteristics, engine mounts, and gaps between points in the vehicle and the barrier or between interior points in the model such as between the engine and fire wall.

Triangular membrane elements are used to represent the sheet metal, which is quite stiff in its plane but has little out-of-plane stiffness. When membrane elements are used, only the three translational degrees of freedom are retained in the solution for those grid points that connect only to the membrane elements. This method for modeling sheet metal reduces the total number of degrees of freedom at the expense of neglecting minor out-of-plane stiffness. The DYCAST membrane element also requires significantly less computational time at the element level than the shell element. Membrane elements are also used to model the engine, where very stiff elements are used since the main requirement is to model the shape of the engine and its rigid body mass characteristics.

Initial conditions specify that each point in the model is moving at the impact velocity of 48.3 kph (30 mph) except for the initial barrier contact point, which has zero velocity. DYCAST then determines what happens to the vehicle in the process of coming to rest. Although several time integrators are included in the program, only the Newmark-Beta method has been used at Chrysler. Typically the entire crash sequence is over in about 75 milliseconds and convergence requires a time step of .05 milliseconds, resulting in over 1,500 time increments. DYCAST alters the time step automatically depending on rate of convergence. If convergence problems are encountered, the time step is reduced. If several iterations are completed with no convergence problems, the time step is increased. One complete analysis of the mini-van requires about 240 minutes of CPU time and 9.5 hours of real time to complete on the CYBER 205.

DYCAST crash results compare quite favorably with test results from the point of view of energy dissipation, total crush distance and mode, and steering column motion, all of which are important in accessing vehicle crashworthiness. Overall acceleration levels also compare quite reasonably, but specific point acceleration values may be significantly different.

Several new features are being implemented in DYCAST by Grumman Aerospace Corporation which should make the program even more useful in the future. A seat-occupant-restraint model will be an integral part of the program. This will enable the effect of a crash on the occupant to be more easily accessed. Currently Chrysler uses an Occupant Simulation Program developed by the University of Michigan (4) to predict occupant behavior. Improvements are also being made in the static analysis capability of the program, which is used to predict roof crush strength of a vehicle. The current version encounters convergence problems due to local instabilities in a model when doing static analysis.

AERODYNAMIC ANALYSIS

The increased importance of the aerodynamic properties of automobiles has accelerated the search for aerodynamic simulation tools. Because of the large number of calculations required for aerodynamic analysis of complex automotive shapes, scalar computers seemed inadequate for Chrysler's aerodynamic applications. Consequently, after acquiring the McAERO program from NASA (5), Chrysler installed a scalar version on a CYBER 760, vectorized the solution phase of the code, and installed it on the CYBER 205. The vectorization resulted in significant time reductions: a model with 354 panels executed in 158 CPU seconds on the CYBER 760 and only four CPU seconds on the CYBER 205 (nearly a 40:1 improvement). Extrapolating these results to large models with the level of detail required for automotive aerodynamic studies indicates that solutions to problems of this magnitude are not possible on present scalar computers. However, the work to date has proven that with the application of the CYBER 205, numerical aerodynamics can be a useful engineering tool. In preliminary tests, calculated pressure values from McAERO correlated very well with wind tunnel test values of a car-like shape.

REFERENCES

(1) **GCSNAST** is a Chrysler-developed pre- and post-processing program for finite element analysis.

(2) **MOVIE·BYU** is a general-purpose graphics display program developed at Brigham Young University in Provo, Utah.

(3) **DYCAST/GAC** User's Manual. Bethpage, NY: Grumman Aerospace Corporation, n.d.

(4) Bruce M. Bowman, Robert O. Bennett, and D. Hurley Robbins, "MVMA Two-Dimensional Crash Victim Simulation, Version 4", Final Report to the Highway Safety Research Institute.

(5) **McAERO** is an aerodynamic analysis program developed by McDonnell-Douglas Corporation for the National Aeronautics and Space Administration, U.S.A.

Session 4 Marine Mathematical Modelling

On the Construction of a Versatile Mathematical Model for Marine Simulation

N.E. MIKELIS, A.J.P.S. CLARKE, S.J. ROBERTS AND E.H.A.J. JACKSON

LLOYD'S REGISTER OF SHIPPING,
71 FENCHURCH STREET, LONDON EC3

Summary

A mathematical model consisting of coupled equations for surge, sway, yaw, roll and propeller revolutions has been assembled for the prediction of ship manoeuvrability at the design stage. All input data describing the ship can be obtained without resource to model experiments, making the use of the method inexpensive and quick. The mathematical model reproduces full-scale manoeuvres with satisfactory accuracy and it is thus an ideal tool for Marine Simulation.

Introduction

Recent work at the International Maritime Organisation aims towards legislation requiring that the designer evaluates [1,2] the manoeuvring behaviour of his design. This has produced growing interest in methods for the calculation and simulation of ship manoeuvring behaviour. Furthermore, in order that a prediction method may be used effectively at the different iterations of ship design, it must not rely on data obtained from model scale experiments since the latter are time consuming and expensive.

With this background this paper presents results of an effort [3] to assemble a mathematical model which accounts for surge, sway, yaw and roll ship motions, for propeller revolutions and for their change during manoeuvres, for rudder motion and finally for a simplified model of the engine. The method employs two computer programs. The first is a pre-simulation program, which uses readily accessible data on installed power and on geometry of ship, propeller and rudder, to generate the resistance, propulsion, manoeuvring coefficients and other data that are needed by the second program which carries out the simulation.

The parametric expressions employed to generate simulation data are not directly applicable to non-conventional hull forms. A facility has therefore been incorporated which allows the program's user to specify data that may have been obtained experimentally or by purely theoretical means.

The simultaneous equations used to describe the time history of a manoeuvre are then solved at high speed by the computer. Typical speeds of over 500 times faster than real time are obtained on the IBM mainframe computer of Lloyd's Register. Much faster than real time response is also expected from a version of this program, presently under development, for a desk top computer.

Initial comparisons using a well known set of full-scale manoeuvring measurements have instilled confidence in the capability of the model to simulate realistically ship responses.

The adopted equations of motion can model a wide range of ship handling conditions and can be extended with relative ease to simulate environmental effects which have not yet been accounted for. One particularly interesting development, based on the manoeuvrability prediction model, has been implemented at Lloyd's Register to provide a simulator for the study and analysis of safety during ship handling operations. This is the Multi-Ship Manoeuvring Simulator (MSMS) which utilises existing hardware (VAX/11 780) and has the following capabilities.

(i) Radar view graphics, with viewing radius that can be chosen at will by the user. Also perspective views are presently under development.

(ii) Interactive input of rudder and engine commands.

(iii) Real time or fast simulation.

(iv) Multi-ship simulation: From one up to five ships can be operated simultaneously, and in proximity to each other.

As this simulator is relatively inexpensive and simple to operate it is expected that in the future it will be used widely for studies such as:

(a) The behaviour of a ship convoy in a canal or in open waters;

(b) Aspects of port safety whereby either proposed port developments can be evaluated for safe manoeuvring operations, or alternatively the use of existing ports by larger or less manoeuvrable ships can be examined;

(c) The effectiveness of a Vessel Traffic System can be evaluated by simulation of traffic 'before and after' the proposed installation;

(d) New ship manoeuvres can be designed and evaluated using simulation. The purpose of such studies would be to produce new standard manoeuvres from which clearer information can be drawn on ship performance;

(e) Conditions leading to collisions or rammings can be reproduced and useful lessons may be learned, as for example, on alternative actions that might be more effective in avoiding an accident.

It is thought that the generality of the manoeuvrability prediction program and the ease with which data can be generated for new ships, could make its adoption very attractive to full mission simulators.

Equations of Motion

To describe a ship's motion, use is made of rigid body dynamics. From Newton's second law the force and moment acting on the ship are equal to the rates of change of momentum and angular momentum respectively. In general these equations describe motions in the six degrees of freedom [4]. It is customary, however, in manoeuvring studies, to reduce the problem by considering only the motions in the horizontal plane, i.e. surge (X force, u velocity), sway (Y, v) and yaw (N, r). In most cases this is a justifiable assumption, although there are some ship types, e.g. fast container ships, that develop considerable heel when turning at speed. For such cases the roll equation (K,p) must be coupled with the horizontal plane equations.

To facilitate the expression of hydrostatic and hydrodynamic forces, a system of body axes is employed (Cxyz on Fig. 1) that is fixed on the ship and is moving relative to the inertial axes OX_0Y_0 with velocity equal to the ship's velocities, u, v, r and $p = \dot{\phi}$. The ship's centre of gravity G is at a distance $(x_G, 0, z_G)$ from the origin C. For convenience, the origin is located amidships at the intersection of the water and the centre plane.

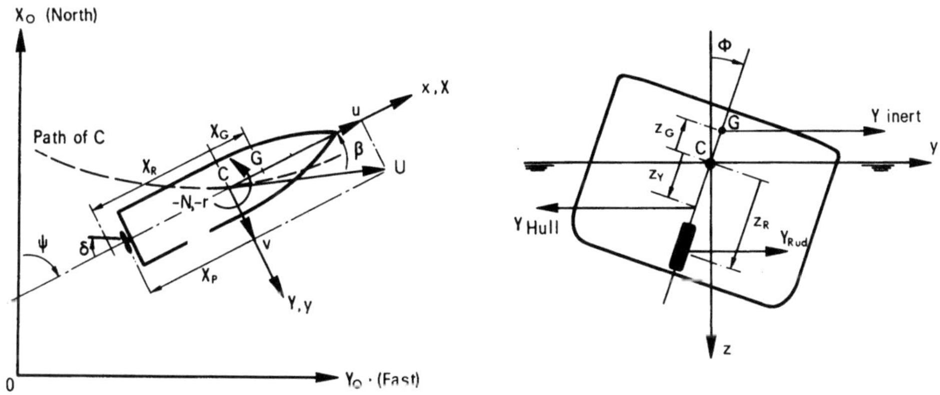

Fig. 1 Coordinate System

From rigid body dynamics the equations of motion (see reference [8] for example) are:

$$
\begin{aligned}
\text{Surge} &: m(\dot{u} - rv - x_G r^2 + z_G pr) &&= X_H + X_P + X_R + X_0 \\
\text{Sway} &: m(\dot{v} + ru + x_G \dot{r} - z_G \dot{p}) &&= Y_H + Y_P + Y_R + Y_0 \\
\text{Yaw} &: I_z \dot{r} + m x_G (\dot{v} + ru) &&= N_H + N_P + N_R + N_0 \\
\text{Roll} &: I_x \dot{p} - m z_G (\dot{v} + ru) &&= K_H + K_{Pr} + K_R + K_0
\end{aligned}
\qquad (1a)
$$

also, the propeller revolutions ,n, are governed by:

$$\text{Propeller: } I_{prop}\, \dot{n} = Q_{Eng} - Q_{Loss} - Q_P \qquad (1b)$$

The terms on the right side of eq. (1a) represent forces and moments due to hydrodynamic reactions on the ship's hull, due to propeller, rudder and other disturbances. The last can be set to represent a variety of actions, such as the effect of wind, of current, the effect of suction experienced by a ship near a canal wall or when overtaking another ship, the use of bow thrusters etc. In these equations the ship's mass is m, its moments of inertia about Cz and Cx are denoted by I_z and I_x, Q_P is the propeller torque, and a dot over a symbol represents time derivative. Next we consider briefly the evaluation of the terms on the right hand side of eq. (1).

The hydrodynamic reactions on the hull can be expressed in a variety of ways, accounting and neglecting of different effects, and thus leading to a number of manoeuvring theories. The classical treatment of Taylor expansions of perturbations around the equilibrium steady forward motion [4] is adopted here [3]. The success of the model relies on the introduction of non-linear terms, expressed parametrically by Inoue et al [5]. The complete expressions for hull forces and moments are given by eq. (2). These are a further development of those published earlier [6] and are not dissimilar from those used by other researchers [7-9].

$$
\begin{aligned}
X_H &= X_{\dot{u}}\dot{u} - Y_{\dot{v}}vr - \frac{u}{|u|}Y_{\dot{r}}r^2 + X_{vr}vr + X(u) \\
Y_H &= Y_{\dot{v}}\dot{v} + Y_{\dot{r}}\dot{r} + Y_v vU + \frac{u}{|u|}Y_r rU + Y_{vv}v\,|v| + Y_{vr}v\,|r| + \frac{u}{|u|}Y_{rr}r\,|r| \\
N_H &= N_{\dot{r}}\dot{r} + N_{\dot{v}}\dot{v} + N_r rU + \frac{u}{|u|}N_v vU + N_{rr}\,r|r| + \frac{u}{|u|}N_{rrv}\frac{rrv}{U} + \\
&\quad N_{vvr}\frac{vvr}{U} + \frac{u}{|u|}N_{\phi}\,\phi U^2 + \frac{u}{|u|}N_{v\phi}\,v|\phi|U + N_{r\phi}r|\phi|U \qquad (2) \\
K_H &= K_{\dot{P}}\dot{P} + K_P P + K_{\phi}\sin\phi - z_Y\,Y_H
\end{aligned}
$$

The multiplier $u/|u|$ on some of the terms here serves to orientate correctly the force and moment terms during backing motion. The choice of which terms are affected is made on physical and symmetry grounds. The ship's speed U, where $U = \sqrt{u^2 + v^2}$, multiplies some of the terms in eq. (2) since the equations are developed in their dimensional form.

The three terms on the yaw moment expression which couple it with roll are taken from [10]. The lever z_Y represents the centre of application of the hydrodynamic force Y_H which is assumed here to act through a point 40 percent of the draught below the waterline.

The hydrodynamic coefficients appearing in eq. (2) can be obtained from a variety of sources e.g. [5,9-12]. The term X(u) in eq. (2) refers to the ship's resistance in a straight course. This is estimated here by the Holtrop and Mennen [13] expressions. For the simulation of manoeuvring motions, a table of resistance values against speed is produced and is then interrogated at each time step of the computation. The same estimation method also yields the wake fraction and thrust deduction which are used in the evaluation of propeller and rudder forces. For reversing motions, an increase in the resistance is derived from data of ref. [14] while the same source also provides a means to estimate the propulsion coefficients when operating outside the first quadrant. The term 'quadrant' refers to each of the four combinations of direction of ship's axial speed and of propeller revolutions (e.g. see fig. 2). The effect of propeller loading on thrust deduction is accounted for by a linear expression fitted to published experimental data , while the variation of wake fraction, w, with lateral motion of the ship is expressed using Hirano's data [8].

Propulsion is modelled by

$$
\begin{aligned}
X_p &= (1-t)\,T \qquad (3) \\
Q_p &= Q
\end{aligned}
$$

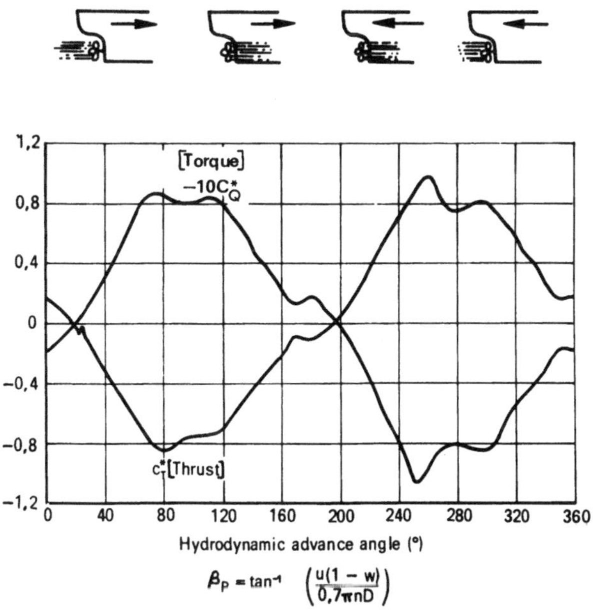

Fig. 2 Interpolated four quadrant thrust and torque coefficients for propeller B5-68 (0.715)

The remaining components of propeller induced force and moment are ignored at this stage.Two sources of data are used to calculate instantaneous values of thrust and torque during a simulation. The first is a polynomial expression obtained from ref. [15], which provides separate polynomials for four and for five bladed propellers. The polynomials are in terms of Blade Area Ratio (BAR), Propeller Pitch to Diameter ratio (P/D) and advance coefficient (J = u(1-w)/nD). The polynomial expressions are employed for moderately small values of J in the first quadrant of operation. When outside this range, use is made of the four quadrant data of the B-Screw Series ([15] with corrections from [16]). These data are Fourier Series descriptions of thrust and torque coefficients for a complete cycle of hydrodynamic advance angle (e.g. see fig. 2). The published data cover the performance of 14 propellers which form three axes (P/D, BAR, number of blades), enabling interpolation to derive data for any intermediate propeller. In the manoeuvring model described in this paper, the Fourier Series format is transformed to a dense enough look-up table prior to entering the simulation. Fig. 2 is a typical example of interpolated four quadrant data for the B-Screw Series equivalent propeller of the VLCC Esso Osaka [17].

For the engine, a simplified model has been adopted whereby to each telegraph setting corresponds a torque value Q_{Eng}. In a simulation when the telegraph setting is altered, there is a built in delay until the new torque is fully applied.

The rudder induced forces and moments and their interactions with the hull are modelled following Japanese researchers, e.g. [7-9]. Additional logic is introduced here to cover operations in four quadrants, as discussed in [3].

To model the effects of wind, waves and bow thrusters use can be made of published work in the open literature to derive expressions for X_0, Y_0 etc in eq. (1a). The complete expressions for modelling current are given in [3]. The current may be unidirectional or it can have gradients in the horizontal plane and/or in time.

When all the terms are assembled in the equations of motion (1a) and (1b), those containing accelerations are grouped on one side. The resulting small system can be solved algebraically so that each of the new equations is an uncoupled expression for the acceleration in each degree of freedom. The computation then consists of a simple time march which provides the velocities in the next time step. The transformations of velocities from body to inertial axes followed by the appropriate integrations finally yield the ship's trajectory.

Verification of the Mathematical Model

Extensive full-scale manoeuvring tests have been carried out [17] for the 278,000 dwt tanker Esso Osaka in deep and in shallow waters. The well documented full-scale measurements are used here for a verification of the deep water mathematical model described in the previous section.

Figure 3 shows graphically for the Esso Osaka in the steady straight motion, the relation between forward speed and RPM, at a range of engine torque settings. Continuous lines on fig. 3 represent results from a number of simulations. Marked on the same figure are points representing speed and RPM settings taken from the initial conditions (steady state) of the full-scale trials. The agreement here is most gratifying and instils confidence in the prediction techniques used to model resistance and propulsion.

A variety of manoeuvres are used to assess a ship's handling behaviour, as discussed in [1]. Two types of manoeuvre are simulated here, turning circles and zig-zags. The published full-scale measurements incorporate corrections to allow for the presence of current, which is assumed to come from a constant direction. However, records [17] of the Esso Osaka trials show a considerable variation in the speed and direction of current in both horizontal and vertical planes. The approximation of unidirectional current appears to have introduced some errors in the published 'corrected' measurements. For this reason further comparisons are planned in the near future using model experimental data, free from these uncertainties.

In the Esso Osaka trials an RPM feedback control kept revolutions constant during turning and zig-zag manoeuvres. Constant RPM has also been simulated in the mathematical model by a simple device for bypassing the RPM equation, namely by artificially increasing the inertia of the propeller to an extremely large value.

Figs. 4 and 5 show comparisons for a 35 degree rudder to port turn, with an initial speed of 7.7 knots. The computed and measured motion parameters in fig. 4, with the only exception of sway velocity, are in very good agreement. The computed curve of sway velocity (at the ship's

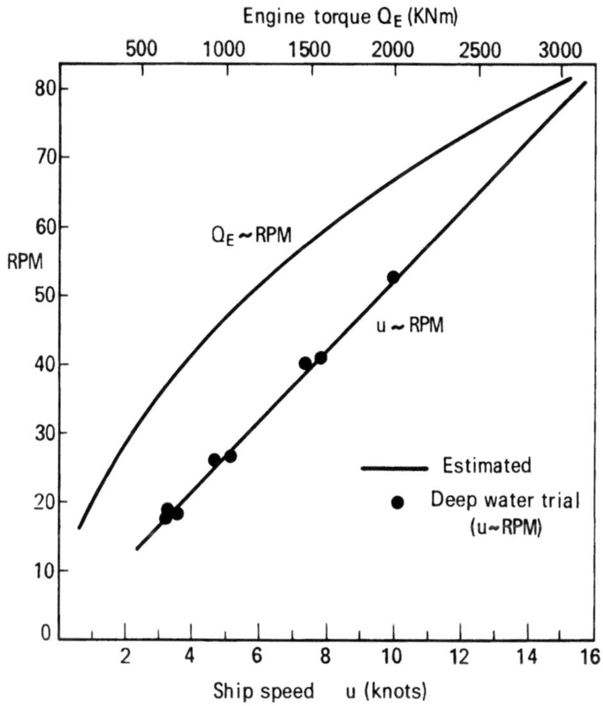

Fig. 3 Esso Osaka: Engine Torque, RPM and ship speed relations

centre of gravity) shows the expected behaviour, whereby a steady state value is reached after the initial disturbance. On the other hand the measured sway velocity (and thus drift angle) shows an oscillation, which interestingly appears to have a period equal to the time it takes the ship to complete a full turn. The almost zero value it reaches around 1300 secs is contrary to the way ships turn, and thus it is suggested that the experimental plot for sway velocity has not been properly corrected for the effect of current. The simulated path of the ship's centre of gravity for the 35 degree turn to port is depicted on fig. 5 as a dot every 50 seconds, while the recorded path of the centre of gravity at the trial is shown by a continuous line. Also shown on fig. 5 is the time in seconds to reach 90, 180, 270, and 360 degrees of turn for the trial and simulation. The agreement appears to be very reasonable up to 270 degrees of turn. After that the experimental path drifts increasingly to the right of the figure in a spiral manner inconsistent with a steady state turn, probably due to an inadequate correction method for the effect of current.

Figures 6 and 7 compare the simulated and measured turning circle at 10 knots with the rudder at 35 degrees to starboard. It is seen from fig. 6 that the agreement for the motion parameters is excellent, with the usual exception of the sway velocity where the measured curve shows again an oscillatory behaviour with a period corresponding roughly to a complete turn. Figure 7 shows the simulated path at 50 second intervals and also

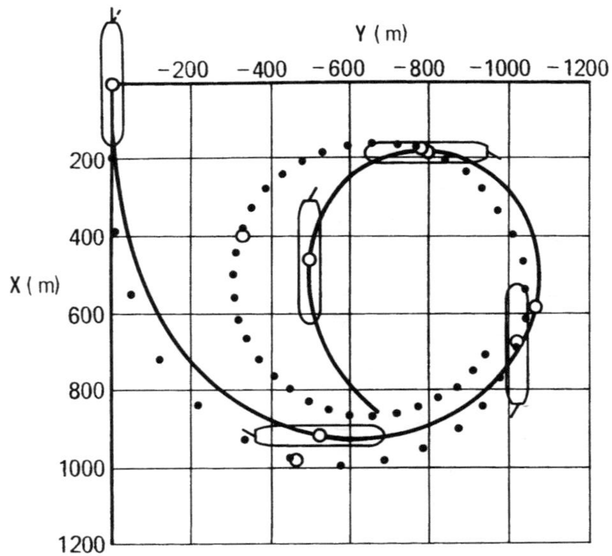

ψ (degrees)	Time (sec)	
	Full scale	Simulated
90	340	352
180	675	707
270	1080	1140
360	1510	1605

Fig.5 Path during the turning circle of fig.4.

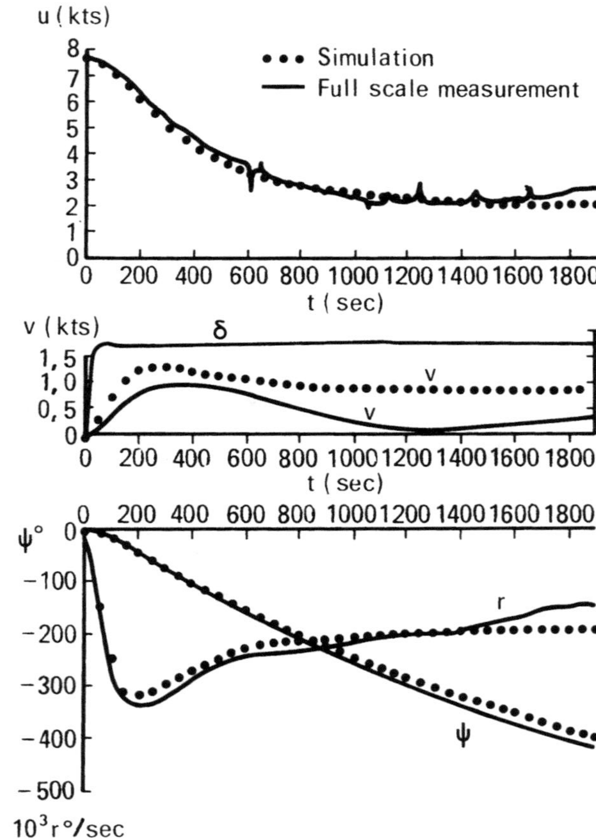

Fig.4 Esso Osaka: Turning circle in deep water (rudder 35° to port, u_o = 7.7 kts)

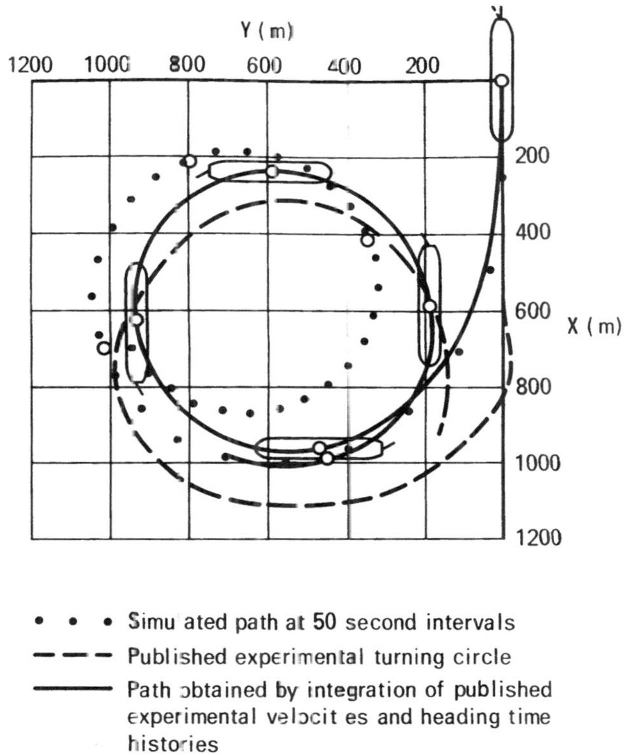

ψ (degrees)	Time (sec)	
	Full scale	Simulated
90	262	270
180	512	534
270	820	858
360	1150	1209

Fig.7 Path during the turning circle of fig.6

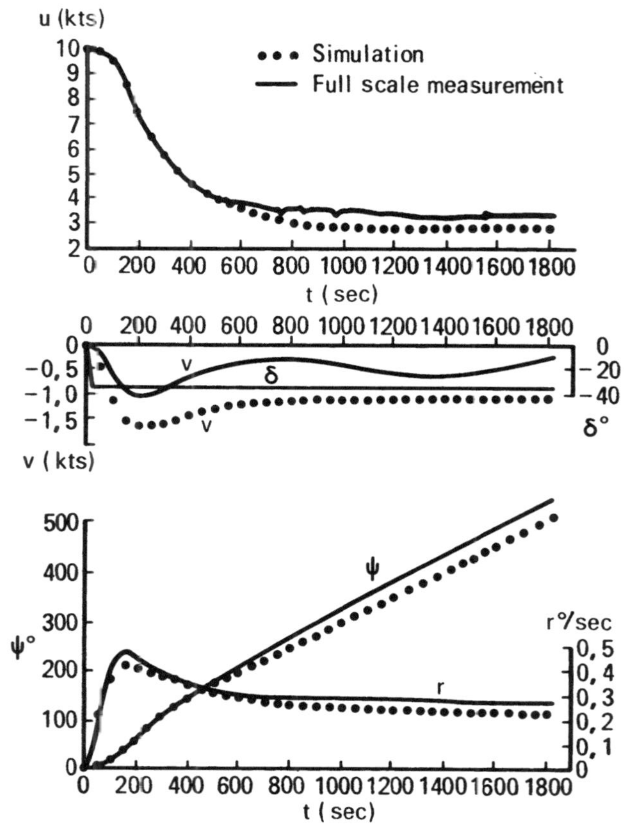

Fig.6 Esso Osaka: Turning circle in deep water (rudder 35° to starboard, u_0=10 kts)

two different curves for the measured circle. Of these, the dashed curve is obtained directly from the report of the trial [17] and is supposed to have been corrected for current effects. No correlation could be obtained between this and the simulated curve however. Also, the published curve is dissimilar to the expected turning circle path (e.g. compare initial stages of measured turns in figs. 5 and 7) and it was therefore checked for consistency with the published motion parameters by an appropriate integration and transformation of the u,v and ψ time histories from fig 6. The result is shown on fig. 7 as a continuous curve, which demonstrates a much closer agreement with the theoretical simulation. It is of interest to note that a very similar turning diameter was obtained from the two manoeuvres in figs. 5 and 7, which confirms expectations for turning at low speeds.

Figures 8 and 9 show comparisons for a 10/10 and a 20/20 zig-zag test respectively. The agreement between full-scale measurement and simulation in both tests is satisfactory and now a much better correlation is demonstrated between the sway velocity time histories. The reason for this better agreement, compared to that in the turning circles, may be attributed to the lesser effect of current on a zig-zag manoeuvre where the average ship's heading remains constant.

The Multi-Ship Manoeuvring Simulator (MSMS)

The simulator MSMS is a single computer program that is able to calculate manoeuvres for one or more ships and to display them in either real or fast time. Each simulated ship occupies two terminals, one for the graphic display of a radar view with own-ship at the centre of the screen and the second one for alphanumeric displays. The latter's screen is divided in two areas, the lower part allowing the input of commands for engine telegraph setting, for rudder orders and for viewing mode. Presently the viewing mode controls the radius of radar while work is underway to also allow for perspective views. The upper part of the alphanumeric display provides performance information, i.e. time, speed, RPM, roll angle, heading and turning rate.

Before starting the simulation a front end program allows the user to specify the type of simulation (real time, fast time or rerun), the output required and the harbour where the simulation will be staged. Also at this stage the user allocates different ships to different pairs of terminals. The main program then dedicates all the required terminals and asks which of the available scenarios (i.e. initial position, heading, telegraph setting etc) is to be assumed for each ship. The main program (driver) runs the simulation by looping round for each ship executing the following logic:

(a) If the user is keying a command then no displaying activity should be done for this ship in order to obtain maximum response.

(b) If no command is being entered then if the simulation is ahead of real time, no further action is taken on this ship and the program proceeds to the next ship. If on the other hand the simulation is lagging behind real or target time then the program performs a sufficient number of time step calculations to catch up with current time.

(c) If computation is ahead of real time by a small amount, then the alphanumeric terminal is updated.

(d) If a suitable number of updates have taken place on the alphanumerical terminal then the graphic's display is updated.

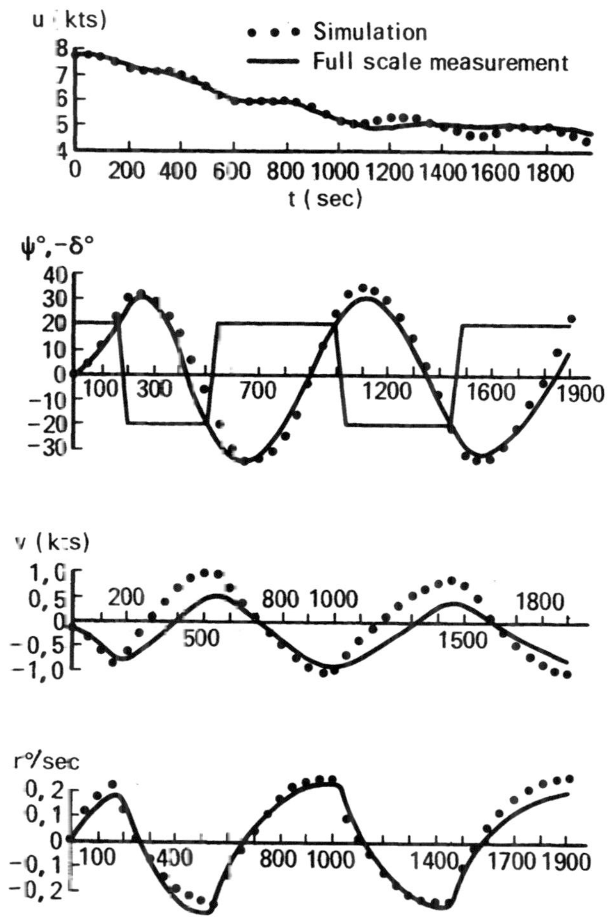

Fig.9 Esso Osaka: 20°/20° zig-zag test (u_o = 7.8 knots).

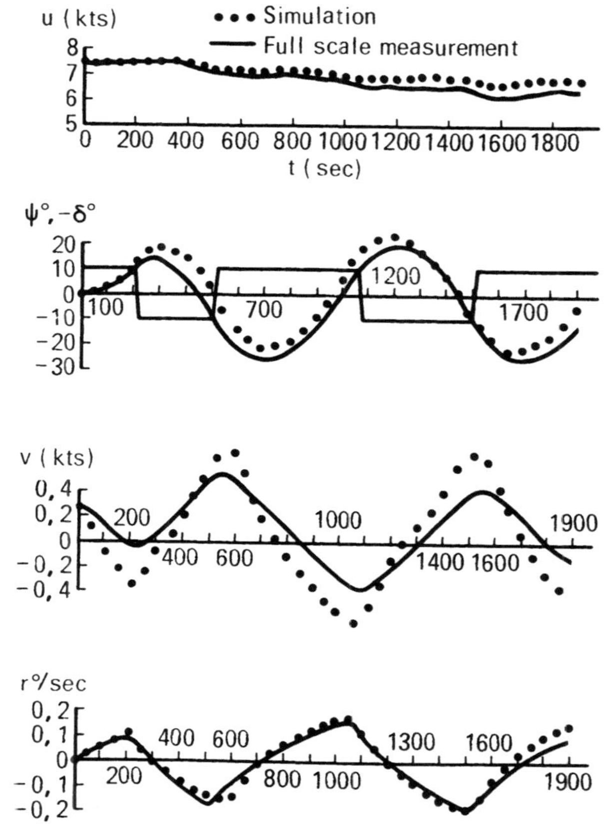

Fig.8 Esso Osaka: 10°/10° zig-zag test (u_o = 7.5 knots).

All graphics employ the 'Template' system (MEGATEK Corp.) using a work station approach were graphics are directed to one of a number of preselected display terminals. The response is dependent on the number of other users on the system, unlike a dedicated computer. Segment double-buffering is used to ensure minimum time intervals between consecutive displays, i.e. the last segment is held on the screen until the next one is built and is ready to be displayed. Compared with the radar view there are more complications in a perspective view (hidden line/face removal and the associated computation time; determination of focus point of the eye and of a realistic viewing frustrum). An investigation is presently being carried out for the optimum display which does not degrade performance of the simulator.

Conclusions

The mathematical model and computer programs developed at Lloyd's Register for predicting and simulating ship manoeuvring responses have been discussed. The ease of generating the input data required by these programs and the economy of operating them has also been pointed out.

Comparisons between computer predictions and full-scale deep water measurements demonstrate satisfactory agreement. Further comparisons with trial data for other ship types and with model scale measurements are planned for the near future. The mathematical model is also applicable to shallow water cases, provided that the necessary hydrodynamic data are available.

Numerous applications of this work are envisaged. In particular it is expected that the Multi-Ship Manoeuvring Simulator, will prove to be invaluable in studies of Maritime Safety.

References

1. IMO, "Interim Guidelines for Estimating Manoeuvring Performance in Ship Design", MSC/Circ. 389, 10th January 1985.
2. IMO, "Manoeuvrability of Ships - Review of Resolution A.209 (VII)", Report of the Ad Hoc Working Group, DE 28/WP.8, 17th January 1985.
3. Mikelis, N.E., "A Procedure for the Prediction of Ship Manoeuvring Response for Initial Design", Int. Conf. on Computer Appl. in Automation of Shipyard Operation and Ship Design, North Holland Publishing Co., Trieste, September 1985.
4. Abkowitz, M.A., "Lectures on Ship Hydrodynamics - Steering and Manoeuvrability", Hydro-og Aerodynamisk Laboratorium, Lyngby Denmark, Report No. Hy-5, 1964.
5. Inoue, S., Hirano, M. and Kijima, K., "Hydrodynamic Derivatives on Ship Manoeuvring", Inter. Shipbuilding Prog., Vol. 28, No. 321, 1981, pp112-125.
6. Mikelis, N.E. and Taylor, K.V., "Towards a Design Tool for Ship Manoeuvreability", Inter. Maritime Association of East Mediterranean, 3rd Inter. Congress, Athens 84, 1984, pp239-249.
7. Ogawa, A. and Kasai, H., "On the Mathematical Model of Manoeuvring Motion of Ships", Int. Shipbuilding Prog. Vol. 25, No. 292, December 1978, pp306-319.

8. Hirano, M., "A Practical Calculation Method of Ship Manoeuvring Motion at Initial Design Stage", The Society of Naval Arch. Japan, Naval Architecture and Ocean Engineering, Vol. 19, 1981, pp68-80.
9. Kose, K., "Prospect of New Generation of Manoeuvering Mathematical Model", Nineth Meeting of the International Marine Simulator Forum at University of Wales.
10. Hirano, M. and Takashina J., "A Calculation of Ship Turning Motion Taking Coupling Effect due to Heel into Consideration", Trans. Society of Naval Architects of West Japan, No. 59, 1980, pp71-81.
11. "Design Information on the Ship Manoeuvrability", Part 1 and Part 2, CMO Project 84 B.2.2, NSMB, December 1984.
12. Mikelis, N.E., "Data for the Evaluation of the Acceleration Coefficients for Tankers Manoeuvring in Shallow and Deep Waters", Inter. Shipbuilding Prog., Vol. 29, No. 340, 1982, pp334-342.
13. Holtrop, J. and Mennen, G.G.J., "An approximate Power Prediction Method", Inter. Shipbuilding Prog., Vol. 29, No. 335, 1982, pp166-170.
14. Harvald, Sv. Aa., "Wake and Thrust Deduction at Extreme Propeller Loadings for a Ship Running in Shallow Water", Trans. RINA, 1976, pp213-236.
15. van Lammeren, W.P.A. et al, "The Wageningen B-Screw Series", Trans. SNAME, Vol. 77, 1969, pp269-317.
16. "Vier-Kwadrant Vrijvarende-Schroef-Karakteristieken Voor B-Serie Schroeven. Fourier-Reeks Ontwikkeling en Operationeel Gebruik", MARIN, Rapport No. 60482-1-MS, Mei 1984, p.30.
17. Crane, C.L. Jr., "Maneuvering Trials of the 278,000 dwt Esso Osaka in Shallow and Deep Waters", Exxon International Co. Report No. EII.4TM.79, 1979, p.174.

Direction of International Joint Effort for Development of Mathematical Models and Ship Performance Data for Marine Simulation Applications

JOSEPH J. PUGLISI, Managing Director CAORF
JOHN S. CASE, CAORF Engineering, Maintenance, and Technical Support Manager
WEI-YUAN HWANG, Project Manager, CAORF Research Staff

CAORF, National Maritime Research Center, Kings Point, New York 11024

This paper discusses international joint efforts to promote a more efficient use of the valuable and costly inventory of ship maneuvering data of different test and simulation facilities. A new concept is presented which allows a simulation facility to more effectively implement a specific ship model based on a standardized data package provided from another facility that has a validated mathematical model in operation. The concept resolves problems with math commonality and significantly reduces costs by minimizing duplication of effort.

INTRODUCTION

Valid ship maneuvering simulation for research and/or training requires ship maneuvering performance simulation to a level of fidelity appropriate for the applied project. To appropriately support research/training needs (schedule and cost) it is necessary to obtain easily at low cost performance data suitable for development of the simulation mathematical model. Sources of data include model towing tank tests, real-world maneuvers, subjective opinions from qualified mariners, and ship maneuvering data from other test and simulation facilities. It is also important to increase the "pool" of available data through cooperative efforts between simulation facilities. This includes sources of data, test data (tank and real-world), mathematical coefficients, etc.

The key problems are that it is very costly to obtain new data of the type to support research/training programs. Availability of suitable real-world/tank data is very limited or non-existent. The data is very costly to obtain and is often contaminated with noise. In addition, simulator facility data are generally non-transferable unless they use common mathematical models.

The data must cover many regimes (shallow water, slow speed, high drift angle, docking, etc.) and does not really exist. Tank data are very limited and not normally available in published literature. Most data generated are normally for resistance/propulsion purposes. Also there is limited awareness of what tank data exists. Tank facilities may have data in file from previous work but unawareness makes it unavailable to the simulation community.

Tank data for meeting ship/bank interaction effects are extremely scarce. Shallow water data are extremely limited. Obtaining tank data for ships operating in shallow confined water can be a very costly task. Interaction effects are difficult to measure, very costly and require special test setups that some towing tank facilities cannot implement. As a result these data are extremely limited or non-existent. Real-world data are also limited and consist mostly of sea trial data (deep/open waters). Data for the ESSO OSAKA are the only good source of extensive full-scale shallow water performance available.

Simulator facility data are generally non-transferable. Mathematical models are typically different making coefficient data non-compatible. Tank data are not easy to implement. The analysis requires significant effort in order to eliminate scatter and noise contamination. Tank and real-world data often requires some type of filtering in order to make it suitable for subsequent development efforts. The tank data used by one facility to implement ship performance must be fully reprocessed by another facility. Resultant performance depends upon data interpretation. Final results can be dissimilar. The validation process applied to certain proper simulation performance must be duplicated when considering another facility. The process may not be duplicated.

In the past few years, there has been an international recognition and joint effort to promote a more efficient use of the valuable and costly inventory of ship maneuvering data of different simulation facilities. The goal is to make the data not only available, but also to be technically exchangeable and comparible between different mathematical models and simulation facilities. Activities currently underway that should minimize costs and minimize data collection efforts include the Joint International Maneuvering Program (JIMP), Reference 1, development of the Marine Coefficient Identification System (MARCIS), Reference 2, and the efforts of the Society of Naval Architects and Marine Engineers (SNAME) H10 Panel (Ship Controllability) and the International Marine Simulator Forum (IMSF).

JIMP was formed to perform cooperative full-scale trial/model test correlations over the complete range of ship designs and still water environments in order to determine model test scale effect factors and physical inconsistencies and to extend the range(s) of mathematical models to include docking/undocking. The U.S. Maritime Administration (MARAD) is currently funding the development of an instrumentation package designed to acquire coefficient data via systems identification during normal operation of a vessel. The SNAME H10 Panel has projects involving hydrodynamic coefficient reporting, developing a primer on the use of mathematical models and defining a comprehensive sea trials agenda to support coefficient development. The IMSF efforts and the Marine Simulation 1984 MARSIM mathematical workshop, Reference 3, have been directed towards defining means of transferring simulation performance between facilities. The key consideration has been the concept of a universal mathematical model that would allow for coefficient commonality between simulation facilities.

This paper discusses these efforts and their impact on the simulation problem and proposes a new concept that was formed because of the continuing problem associated with a simulation facility being able to effectively implement a specific ship model for research/training purpose based on data from another simulation facility that has a validated mathematical model in operation.

ON-GOING ACTIVITIES

The first workshop, entitled "Workshop for Development of an International Joint Ship Maneuvering Program", was held in June 1981 under the sponsorship of the U. S. Maritime Administration, The Society of Naval Architects and Marine Engineers, and Webb Institute of Naval Architecture. The purpose of that particular conference was to propose and plan a "Joint Interantional Maneuvering Program (JIMP)." The JIMP would:

- organize the performance of full-scale trials and scale model tests across a full range of ship designs and environmental conditions
- determine the scale effect factors which must be considered in the application of model tests to the developement and validation of general purpose mathematical maneuvering models
- be responsible for guiding the extension of existing mathematical models to include docking and undocking maneuvers

The ultimate aim of the "JIMP" would be to improve and expand the mathematical models used in ship maneuvering simulations. The "Workshop for Development of an International Joint Ship Maneuvering Program" joined together approximately 50 of the key people from North America, Europe and Asia who were involved in mathematical modelling for marine simulation applications. The Joint International Maneuvering Program was conceived because of the limited number of experts and size of the field and recognition that to make significant progress, extensive and carefully performed full-scale trials, model testing, and theoretical analysis must be performed. Such work is expensive and could possibly be pursued on a cooperative basis where resources are shared for the common advancement of all.

Current JIMP activities involve towing tank facilities worldwide performing ESSO OSAKA tests using the same tank models. Test results are being evaluated based upon the ESSO OSAKA test data to improve the state-of-the-art in tank model testing by reducing test errors through a better understanding of the impact of model scale size and system linearities and improving predictability of full-scale performance from model data.

The MARCIS instrumentation package was designed to allow installation on a "ship of opportunity" and, using System Identification techniques, would generate a mathematical model that would describe the ship's behavior based on data acquired during essentially normal operation. The system is currently in its final stages of development and has been installed on several ships including a U. S. Coast Guard Cutter. Performance evaluation is currently underway.

In June 1984, the IMSF sponsored a "Workshop in Mathematical Modelling of Ships Maneuvering Behaviors" that was held with the objective of determining the practicality of a standardized mathematical model to be used in training and research simulators. Another objective involved defining a plan of action to set up the specifications and define a new structure of mathematical models to be used in marine simulation, and to plan the funding of such work. Four workshop groups met to discuss the following: areas of application, the model structure, determination of coefficients, and the organization and funding of the effort. Although preliminary guidelines were formulated, some participants indicated a reluctance in using a universal model, primarily because of a desire to continue using their current models. They also emphasized the need to share, between facilities, the towing tank and real-world data that was used during the coefficient development process.

The IMSF is also planning to address the issue of data availability to all simulator facilities by developing a bibliography on towing tank testing and a bibliography on ship types that have been implemented on ship maneuvering simulators worldwide.

The on-going activities will provide some measure of support but will not resolve all of the issues:

- JIMP effort will require significant time to yield helpful inputs. Applicability is primarily towards improving simulation fidelity, not ease of implementation.
- MARCIS package in final development stage.

The MARCIS package derives ship performance based on System Identification. The ship maneuvering simulation community is not currently working with this modelling approach. Although system identification has the potential to yield high fidelity performance, costs can be extremely high because of real-world maneuvering requirements for full regime simulation. The current hardware configuration is not portable in the true sense of the word. It currently does not apply to the "ship of opportunity" concept. Also, package installation/checkout may be costly and time consuming. An effort is underway to improve the package for ease of operational transportability.

The concept of a universal mathematical model is not being readily accepted in the industry. Users appear to be reluctant to implement a new model because of:

- Satisfaction with models currently in use.
- Differing opinions on what the new model structure, fidelity, etc. should be.
- Difficulty in developing the model (who will do it ?, how will effort be funded ?, model validation, etc.).

The MARSIM mathematical modelling workshop results implied that one practical method to achieve simulator facility commonality/transfer of ship performance would be through sharing of source data, not coefficients.

The efforts mentioned above are not directly leading towards reducing effort or costs associated with developing ship maneuvering characteristics. Also, the efforts are not directly leading towards solving the simulation commonality problem. Because of differing mathematical models, facilities still cannot benefit from previous efforts undertaken in developing mathematical model coefficients. Problems still remain:

- Source data still limited or non-existent.
- Because of problems with source data consistancy, noise, etc., sharing of this data does not necessarily yield similar resultant simulator characteristics.
- Working with common source data requires parallel data analysis, model development and validation efforts. Efforts undertaken by one facility are not transferable to another.

CONSIDERATIONS

Certain factors need to be considered:

The ideal situation is where most of the efforts undertaken to get a ship simulation capability at one facility can be transferred to a second facility. This includes source data analysis, implemented performance and validation efforts.

It is desirable to be able to use mathematical models in place. This implies the need to use existing coefficient development process in use for each facility.

It is desirable to use for source data tank-type information which is filtered, noise free, free of scatter and scale effects, etc. Ideally, if this data were implemented on dissimilar models, performance for identical situations and conditions would be comparative. In a recent study, Reference 4, for the Panama Canal Commission on defining criteria for the widening of the Gaillard Cut at the Computer Aided Operations Research Facility (CAORF), tank data was implemented on the two different mathematical models at CAORF and the Swedish State Ship Experimental Institute (SSPA) facility. Comparable results were achieved.

It is highly desirable to have performance validation information for maneuvers exploring the full operational regime in order to ascertain proper implementation.

It is highly desirable to have full source and performance data to insure proper simulation throughout the applicable regime.

It is important to have the source data of the type (rotating arm vs. PPM vs. real-world maneuvering) that is compatible with the recipient facilities coefficient development process.

NEW CONCEPT

Simulation facilities having a working validated mathematical model know its capabilities, limitations and fidelity and know its applicability to research and training. The development of the mathematical model usually involves considerable expense, especially if the objective is to obtain high fidelity performance for research purposes. Its performance takes into account the analysis and filtering of towing tank and real-world source data and any coefficient fine tuning that resulted from the validation effort.

The new concept proposed, involves exercising the mathematical model using standardized test guidelines with the simulation configured in two ways – as a towing tank captured model and as a free running ship. As a captured model, output data could be obtained that would resemble data from towing tank captured model tests. As a free running ship, the output data would resemble data obtained from real-world tests. The data would fully characterize performance for all operating regimes and be significantly more comprehensive and lower cost than source towing tank and real-world data since operation would probably be done off-line.

It is also proposed that the standardized test guidelines be prepared by the IMSF. These data, along with other information discussed below, could be used by another user simulation facility to develop coefficients and validate the implemented ship's performance. CAORF has had SSPS translate the model test data that was generated for the Panama Canal Project into coefficients for Dr. Norrbin's mathematical model. The mathematical model was exercised to provide anticipated maneuvering data for the Golden Dolphin Panamax Class ship. CAORF has done the same for its mathematical model. CAORF plans to exercise the model to generate pseudotank data and compare it to the SSPA source tank data.

CONCEPT ADVANTAGES

Utilizing this concept would:

- Result in a comprehensive data package for use by another user facility in developing coefficients for its own mathematical model.
- Provide resultant performance that would be traceable to original source without requiring independent analysis.

- Provide resultant performance whose validity would be traceable to the original source.
- Minimize development cost. Significant savings would be realized depending upon the effort originally expended to analyze tank/real-world data, develop coefficients and validate performance. Validation effort alone can be a very significant cost factor.

Figure 1 shows activities associated with the source facility. The output consisting of a standard package of information is shown in Figure 2. Information in addition to the pseudotank and performance data is needed by the user facility so that a complete understanding of the source facilities model performance can be realized by the user facility to insure valid application to the intended research/training project.

CONCEPT REQUIREMENTS

Certain conditions must be met for the concept to be a viable approach that will result in low cost implementation of a ship currently in another simulation facilities inventory.

- Pseudotank and performance data
- Source facility's implemented model must be of sufficient fidelity and cover the operational regime of recipient facilities intended research/training application.
- Information package from the source facility must be complete and include:
  - Pseudotank and performance data
  - Source data description and evaluation (accuracy, noise content, data scatter, comprehensiveness).
  - Relative level of fidelity of implemented model.
  - Validation process description.
  - Correlation effectiveness with source data.
  - Effectiveness of coefficient development process.
  - Effectiveness of performance indicies matching.
  - Description of known mathematical model limitations.
  - Source performance information and evaluation.
  - Effectiveness of performance indicies matching.

Since most of the facilities or supporting manufacturers have developed processors that require certain source data, the concept presented herein provides alternative methods for generation of the pseudotank and performance data. These processes include Rotating Arm, Planar Motion Mechanism, and System Identification (SI). The set of test maneuvers and the resulting output data generated by the source facility will then be compatible with the users coefficient development process. With regard to the MARCIS SI output, consideration is being given to exercising the SI model, thereby generating the pseudotank and performance data for coefficient generation use at another facility.

IMSF SPECIFICATION

The proposed specification contains test and specifications, as shown in figure 3, that will result in a complete set of pseudotank/real-world test data. The specification provides:

- Format of output data.
- Initial conditions and run duration for pseudotank test runs (both dimensional and non-dimensional).
- Environment test conditions for pseudotank test runs and pseudoperformance indices runs (both dimensional and non-dimensional).
- Description of test set-up.
- Description of test process.
- Description of source data evaluation criteria.
- Description of relative level of fidelity evaluation criteria.
- Description of pseudoperformance data evaluation criteria.
- Description of performance indices matching evaluation criteria.

The specification should be constructed to specify the following:

- Hull Hydrodynamics Tests
  - Captive Model Tests
    - Rotating Arm
    - PPM
- Free Running Tests – for SI Techniques
- Interaction Effects Tests
- Mathematical Model Information
- Tank Data and Full Scale Data Evaluation
- Mathematical Model Validation Information
- Towing Tank Test Information
- Data Specifications
- Standard Performance Tests and Indices
- Full-Scale Ship Specifications
- Towing Tank Model Specifications
- Facility Specification
  - Towing Tank
  - Simulation
  - Report Format and Organization

Facilities requesting data for a given ship type will need to identify that subsection of the IMSF specification that would result in data that would be compatible with their mathematical model development process. It is also desirable that the facility identify the operational regime requirements following the guidelines of the specification. This will insure that forthcoming data will be applicable for the intended research/training project.

CONCEPT IMPLEMENTATION

Certain conditions must be met in order for the proposed concepts to be considered viable and implementable. These include:

- Concept acceptance by the simulation community.
- Source facility willingness to modify their simulation software to accommodate operation as a captured model.
- Willingness of the source facility to generate and release data.
- Development of universally accepted IMSF specification.
- Source simulation facility and operational regime is greater than the needed recipient.
- Completeness and usability of the IMSF specification.
- Trust between source and recipient facilities.

SUMMARY

In order to meet the research and training needs of the maritime community, ship maneuvering simulation facilities must be capable of simulating the maneuvering performance of a given ship type with a level of fidelity consistent with the planned project. The effort and cost associated with implementing a ship on a simulator can be quite considerable and represents a significant technological challenge. Towing tank model tests, real-world ship maneuvering data, subjective evaluation comments from pilots and other knowledgeable mariners and much analysis and evaluation enter into the effort. Fine tuning of coefficients to get the performance "just right" is an important final step that takes into account all data and supporting information. Major problems that must typically be overcome during the process include:

- Towing tank data may not exist. If performance extrapolation from similar ship types is not appropriate than costly tank testing must be undertaken. Existing towing tank data invariably requires a data analysis and filtering process in order to eliminate noise and reduce scatter.
- Similar problems exist with real-world effects, passing ship and bank/channel interaction effects. Data is limited or non-existent and can be difficult or impossible to get, depending upon scenarios and restricted waterway configuration to be simulated.

- Other simulation facilities may have an applicable ship in their inventory but because of mathematical model dissimilarities, the only data considered transferable are the supporting tank and real-world source data. Essentially all efforts, from data analysis through validation must be repeated.

Several ongoing activities have been undertaken in order to address some of these problems. The JIMP program has involved the maritime community worldwide in a cooperative effort to improve the state-of-the-art in captive model testing and simulation mathematical models.

The MARCIS instrumentation package development effort was undertaken to enable acquiring coefficient data via system analysis during essentially normal operation of a vessel for use in simulation work and in building up of a library of mathematical models of vessels in different conditions.

The IMSF has been examining the concept of a universal mathematical model for all facilities in order to resolve the mathematical model commonality problem. Workshops have been held to determine the applicability and acceptability of this concept.

As JIMP goals are achieved, it is expected that an improvement in the quality of ship maneuvering simulation will result. This unfortunately does not address the more critical problems that are associated with the effort and cost to achieve a simulation capability. The MARCIS instrumentation package, when operational as originally conceived as a portable device, may prove an asset under appropriate circumstances – ship data acquired are a candidate for simulation, but real-world maneuvering must be adequate for full regime simulation and simulation facilities must be willing to reprogram to use the new models. The universal mathematical model concept requires a significant development effort and acceptance by all users in order to resolve the mathematical model commonality problem between facilities.

A new concept has been proposed herein that would allow one simulation facility (source facility) transfer to another facility (user facility) information regarding an implemented ship that would minimize and in some cases eliminate the need to duplicate efforts during implementation. Mathematical Model commonality is not required. Information being transferred does not include coefficients for use in another facilities model. The concept basically involves running the simulated ship off-line in both a captured model and free running mode in accordance with specific test guidelines. The resultant data package represents a comprehensive set of pseudotowing tank data and performance indicies/maneuvers that represents the validated maneuvering performance and interaction effects for the subject ship.

The test guidelines, developed under the auspices of the IMSF, would be sufficiently comprehensive to enable source facility generation of data for full-regime characterization and user facility coefficient development, implementation and validation. The guidelines would also identify supplementary information to be provided by the source facility so that the user facility would understand the applicability of the simulated ship to the intended research/training project. In order to test viability of the proposal concept, the guidelines could be exercised for the CAORF mathematical model for the Golden Dolphin Panamax Class Ship. This model is probably the most highly validated mathematical model correlated with full-scale data available in the world today for restricted water operation.

User facility implementation of a source facility's ship using this approach has several key advantages over other concepts:

- Implementation costs are significantly reduced. Source data re-analysis for user coefficient development is not required. Duplication of source facility validation efforts is not required. Validation consists of comparative analysis of performance indicies/maneuvers.
- Implementation time is reduced since effort is less and all work can essentially be done off-line.
- Simulation confidence is higher. If a source facility has a well validated high performance model, the user facility model should exhibit comparable performance.
- Coefficient development process is simplified since source facility data is comprehensive and free of noise and scatter.
- Pseudotank data are compatible with user facility coefficient development process. As an example, source facility model development may be based on PPM type tank data, while user facility may desire rotating arm type data instead. Pseudotank data would be provided in the desired format.

The concepts presented herein can represent a viable solution to the problems associated with mathematical model commonality, transfer of simulated ships characteristics data from one facility to another for implementation, duplication of effort, etc. The maritime community, in particular simulation facilities, need to evaluate these concepts and provide to the authors appropriate comments that can serve as guidelines for continued work.

The concepts presented are based on an analysis of the problems with defining and implementing ship maneuvering characteristics and interaction effects. The authors also solicit comments regarding other considerations that could have a bearing on the overall concept or technical details.

During the following workshop several points should be discussed that will have a key bearing on subsequent work and direction. Discussion points should include:

- Concept visibility
- Criteria for acceptance
- IMSF inspection
  - Structure
  - Details

REFERENCES

1. National Maritime Research Center. Proceedings "Workshop for Development of an Internationa Joint Ship Maneuvering Program (IJSMP)." National Maritime Research Center, Kings Point, New York 1981.
2. Systems Control Inc. "A System for Estimation of Hydrodynamic Coefficients for Full-Scale Ship Tests," National Maritime Research Center, Kings Point, New York Report NMRC-207, 1981
3. MARSIM (1984). Proceedings "Third International Conference on Marine Simulation." Maritime Research Institute Netherlands (MARIN), Rotterdam, The Netherlands, 1984.
4. Puglisi, J.J., D'Amico, A.D., and Van Hoorde, Jr., G."The Use of Simulation at CAORF in Determining Criteria for Increased Throughput of Ship Traffic in the Panama Canal." Proceedings of the Third International Conference on Marine Simulation MARSIM 84. Maritime Research Institute Netherlands (MARIN), Rotterdam, The Netherlands, 1984.

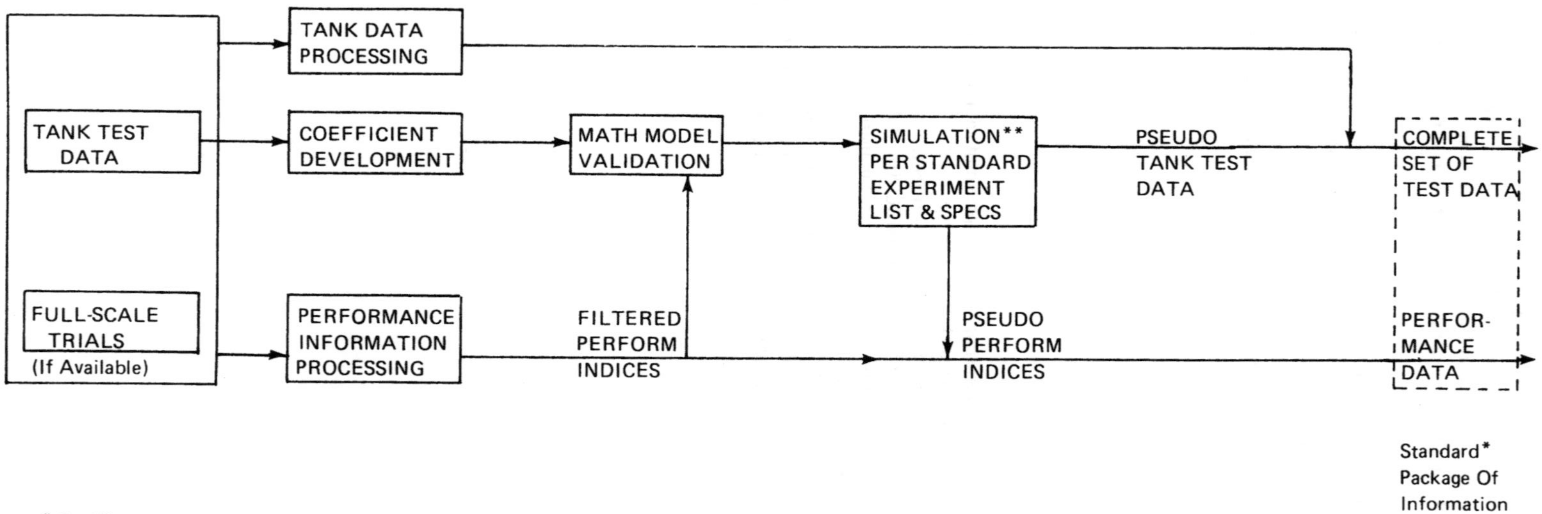

* See Figure 2
* See Figure 3

Figure 1. Activities at Source

- "Complete" Tank Test Data
  - Pseudotank Test Data
  - Raw Tank Test Data
  - Filtered Tank Test Data
- Performance Indicies
  - From Physical Test (Full and Model Test Data)
  - From Mathematical Model Simulation
- Tank Data and Full-Scale Data Evaluation
- Information of Mathematical Model,
  - Structure
  - Parameter Values
  - Accuracy and Fidelity
- Information for the Validation of Mathematical Model
  - Description of Validation Process
  - Criteria of Validity
  - Evidence of Validity
  - Operational Region and Limitations
- Information of Real Tank Test
  - Description of Tank
  - Description of Test Facility
  - Description of Experiment Design

Figure 2. Standard Package of Information Transferred

- Test Specifications
  - List of Requested Tests

Environ
- Deep
- Shallow
  - Open
  - Restricted (Optional)
    - No Passing
      - Calm
      - Wave
    - Passing
      - Meeting
      - Overtaking

Experiment
- Passive
  - Pure Sway
  - Pure Yaw
- Free Running
  - Zig Zag
  - Turning
  - Spiral
  - Crash Stop

Sway-Yaw Coupled
Resistance
Slow Ahead
Slow Astern
Bollard Pull
Bollard Push

  - Initial Condition
  - Duration
  - Range of Test Parameters
- Data Specification
  - Plot Specification
  - Tape Storage Format
  - Unit
- List of Standard Performance Indices
  - Eigen Values
  - Over-Shoot of Z Maneuvers
  - Tactical Diameter, Transfer, Advance
  - Crash Stop Head Reach and Side Reach
  - Interception of S-R curve
- Full-Scale Ship Specification
- Model Specifications
  - Ship — • Bare Hull, With Propeller, With Rudder
  - Restriction (Optional) — • Water Depth, Channel Width, Bank Slope
  - Disturbance (Optional) — • Wave Current, Passing Effect
- Facility Specification
  - Carriage, Tank, Instrument — Towing Tank Facility
  - Computer — Simulation Facility
- Report Format and Organization

Figure 3. Items of Specification and Format

Session 5 CFD for Marine Vehicles

The Computation of Flow Around Ships with Allowance for Free-Surface and Density-Gradient Effects

D. BRIAN SPALDING

Computational Fluid Dynamics Unit
Imperial College of Science and Technology
London, UK

Introduction

The numerical computation of the flow around ships

It has been possible for several years to obtain solutions of the Navier-Stokes equations by numerical means, and only the cost of doing so has delayed large-scale use of the methods for predicting tho flow around surface and submarine vessels.

The cost of making the solutions arises from the fact that large numbers of grid nodes are required for adequately-precise representation of the flow fields; therefore large storage capacity is required of the computer, and the number of arithmetic operations to be performed is very great.
Fortunately, developments in computer hardware are reducing the costs dramatically; and improvements in solution algorithms, for example the adoption of implicit rather than explicit formulations of the equations, have also made a major contribution.

In addition, the greatly improved input-data procedures of some computer codes for simulating fluid-flow processes have opened their use to non-specialists.

Finally, research into turbulence has produced "turbulence models" which, to a sufficient extent, represent the important viscous effects in boundary layers and wakes.

As a consequence, practical interest in the use of numerical methods for predicting flow around ships is growing, as evidenced by the recent SSPA conference (Lindgren H, 1985); and the trend may be expected to accelerate.

The inclusion of free-surface effects

The calculation of the flow of water around surface vessels, and around those which are not too deeply submerged, is complicated by the fact that the surface of the water becomes distorted, in a manner which is not known in advance.

In most calculations which the present writer has seen, this fact is simply ignored: the surface is taken as being plane, and the pressure there is allowed to vary. In reality, of course, the surface pressure must remain atmospheric, a condition which is achieved through a change of surface height by the amount $\delta p/(\rho g)$, where δp is the pressure change to be accommodated, ρ is the density of the water and g is the gravitational acceleration.

Allowance for the surface distortions, ie for the formation of waves, has not in the past been easy. Methods of doing so are reviewed, to some extent, in a recent publication by R W Yeung (1982), which is however confined to "potential" flows, ie to those which are free from viscous effects.

In the present paper, three methods of allowing for distorted-surface effects will be reviewed. It is premature to pronounce upon which is preferable for flow-around-ships simulation; but their existence implies that it is now just as easy to include free-surface effects as to neglect them.

Three methods of analysis

The GALA method

More than ten years ago, the present writer was concerned with the simulation of a postulated accident in a nuclear reactor, in which a bubble of high-pressure sodium vapour expanded within a pool of liquid sodium, causing the upper surface of the latter to strike the roof of the container.

In this problem it was necessary to simulate numerically the motion of both the liquid and the vapour; and the interface distortions were considerable. The practice adopted was therefore to keep the computational grid fixed in position, and to allow the interfaces to cross grid lines.

Some difficulty was encountered in handling numerically the presence of the density discontinuities at the liquid/vapour interfaces, until a re-formulation was arrived at (Spalding, 1974), which can be explained as follows.

The mass-continuity equation can be written as:

$$\frac{\partial \rho}{\partial t} + \text{div}\,(\rho \underline{v}) = 0 \qquad (1)$$

where ρ is the density, t stands for time, and $\underline{v}$ represents the velocity vector. However, it can also be written in the equally-valid alternative form:

$$\frac{D\,(\ell n \rho)}{Dt} + \text{div}\,(\underline{v}) = 0 \qquad (2)$$

where D/Dt represents the operator $\partial/\partial t$ + $\underline{v}$.grad, ie the "substantial derivative", and the density has disappeared from the "div" term.

To each of these differential-equation formulations there corresponds a finite-volume one; and the one corresponding to (2) is much easier to program than that corresponding to (1). The reason is that the evaluation of $\rho \underline{v}$ requires determination of an average density for each cell boundary, which is difficult near vapour˜liquid interfaces; whereas evaluation of $D(\ell n \rho)Dt$ merely requires knowledge of how the densities of identified particles of fluid vary with time, which is easier to come by. Indeed, when the fluids are incompressible, as is effectively the case for flow around ships, equation (2) reduces to:

$$\text{div}\,(\underline{v}) = 0 \qquad (3)$$

which implies that knowledge of the densities is immaterial to the continuity equation, this being identical in form to that for a single uniform-density fluid.

The numerical procedure built upon equation (2) is called GALA for (GAs-Liquid Algorithm); and it has been most extensively exercised by Maxwell (1977), in conjunction with a conceptually-simple particle˜trajectory technique for establishing the location of the gas˜liquid interface at each instant of time. Agreement with experimental data proved to be excellent for phenomena investigated, namely the rise of a bubble through a liquid, the "slumping" of an unsupported mass of liquid, the transient motion of a water wave, and its "breaking" over an obstacle.

The work of Maxwell was conducted with one of CHAM Ltd's early codes for solving the Navier-Stokes equation. However the more-recently-developed PHOENICS code, (Rosten and Spalding, 1985) contains the GALA and particle-tracking options, and can therefore be used for similar calculations.

The "blockage" method

GALA, as just explained, performs calculations over a volume of space which contains (in different parts) both gas and liquid. However, when the density ratio is as great as pertains to water and atmospheric air, the space occupied by the air proves to be of nearly uniform pressure, because of the low inertia of the latter. A uniform-pressure presumption would be quite good enough.

In flow-around-ship problems, the air pressure is not only uniform: it is k n o w n; and this feature permits the employment of alternative methods of solution.

One of these is the "blockage" method, which is easy to use when the computer code is already equipped with the ability to assign multiplying factors for every area and volume of a computational cell, which is true of PHOENICS.

In the blockage method, the equations of motion and continuity are solved only for the denser fluid, i.e. the water in the case of a ship; and the volume occupied by the atmosphere is rendered inaccessible to the water by the setting of the corresponding blockage factors to zero. The computational grid is fixed and extends both above and below the gas~liquid interface.

Computational cells which lie wholly below the interface have blockage factors of unity, except at the upper layer of cells where values in excess of unity are allowed. Those wholly above the interface have blockage factors of zero; and those which enclose parts of it have factors lying between 0.0 and 1.0.

How, it will now be asked, are the values of the blockage factors to be set for the interface cells? The procedure is as follows:-

(1) The interface shape is guessed, e.g. as being planar.

(2) The Navier-Stokes equations are solved, with the result that a pressure distribution is computed for the cells through which the interface passes (i.e. those just below the fully-blocked region).

(3) For interface cells for which the computed pressure is above that prescribed for the atmosphere by the amount δp, the blockage factors are reduced, which means that the implied interface position is raised. For interface cells in which the δp is negative, conversely, the blockages are increased, i.e. the implied interface height is reduced. The height changes are equal to $\delta p/(\rho g)$, so as to conform to the hydrostatic laws.

(4) The Navier-Stokes equations are solved again, with the consequence that a new set of interface-cell pressures is computed, which pressures are usually found to be closer to the values prescribed for the atmosphere.

(5) The process of interface adjustment is continued until the degree of satisfaction of the interface-pressure criterion is deemed sufficient.

This procedure is easy to understand; but its ease of execution depends upon the flexibility of the oomputor code and on the availability to its user of simple means for altering the values of the blockage factors. Fortunately, the PHOENICS code both possesses an efficient Navier-Stokes solver and provides its user with complete control over the blockage factors. It has therefore permitted hitherto-difficult problems to be solved, such as the computation of the shape of a bath-tub vortex (Pericleious, 1985). This method is very suitable for the simulation of flows around ships which are influenced by free-surface effects.

It should be mentioned that the freedom of the PHOENICS-user to vary blockages can be exploited in other ways. Thus, instead of varying only the blockage factors in the vicinity of the interface, he can vary all the factors in a vertical column of cells in the same proportion, thus in effect employing a vertically-distorted grid. This practice has been adopted for computation of the internal-wave phenomena which occur in aluminium smelters; and it could easily be employed for flow-around-ships problems, and to internal waves in gravity-stratified fluids.

<u>The "pseudo-compressibility method</u>

It is well known that there exists a useful analogy between the equations governing the two-dimensional flow of a compressible gas on the one hand and those governing the flow of a shallow layer of water on the other. Specifically, the gas has to be one having a specific-heat ratio of 2.0; and the water layer should lie on a horizontal frictionless surface.

PHOENICS possesses a compressible-flow option; and this, if the specific-heat-ratio is set equal to 2.0, can therefore simulate shallow-water phenomena. What is less obvious, but equally true, is that it can be employed for deep-water phenomena as well, as will now be explained.

The user of PHOENICS is at liberty to calculate density in any way he chooses; he can therefore decide to make the fluid compressible in the upper layer of cells of his computational grid, while retaining a uniform density in all other cells. This practice simulates the way in which the gas-liquid interface moves, in response to changes of pressure influenced by the gravitational field.

This simple practice, it should be mentioned, does not precisely accord with the dynamical laws of deep-water flows; for in shallow-water theory account needs to be taken only of the horizontal forces in the layer whereas in deep-water theory both horizontal and vertical forces are significant. In the compressibility model both forces are linearly dependent on pressure; while for shallow water the horizontal force increases with the square of the layer height whereas the vertical force increases only in direct proportion to the height. This defect can be remedied by further use of the blockage feature.

Examples of the calculation of flow around floating and submerged "vessels".

Problem specification

In order to illustrate the pseudo-compressibility method, the PHOENICS computer code has been applied to the flow around a rectangular object having dimensions:- vertical height 1m; horizontal length 4m; and horizontal width 2m.

This object is supposed to be moving along a canal of width 10m and depth 2.5m at a speed of 1 m/s. Two cases are considered: in the first, the upper surface of the object lies in the plane of the distant water surface; and in the second it lies 0.5m lower. The flow is taken as steady. The calculation domain extends completely across the canal, and from 3m upstream of the object to 8m downstream of it.

Some details of the computations

All the calculations have been performed with uniform grids having 5 divisions in the vertical direction, 15 divisions in the horizontal length direction and 10 divisions in the horizontal width direction. The use of only 750 cells does not, of course, make for very accurate calculations; but they were performed on a

Perkin-Elmer 3220 mini-computer in times of the order of 10 minutes for each run, for reasons of economy; and their accuracy is sufficient for the present limited objective, which is simply to point out some major features of the calculation.

Results will be presented from five distinct runs, namely

1. Floating body, rigid upper surface
2. Floating body, free surface;
3. Submerged body, rigid upper surface;
4. Submerged body, free surface;
5. Floating body, shallow water.

Here the phrase "rigid upper surface" implies that free-surface effects were neglected, as has been the case in most earlier Navier-Stokes solutions; and "shallow water" implies that the canal depth was reduced to 0.5m, so as to equal the vertical dimensions of the object. In the latter case, of course, it was not necessary to solve for the vertical velocity component at all, and the problem became a two-dimensional one.

The results to be shown will consist of line-printer outputs representing contour plots of the computed average pressures in the upper layer of cells. These are indicative of the distortion which would take place in the water surface.

Of course, complete tables of the pressure and of the three velocity components are also produced by PHOENICS; but the contour plots suffice to show how the allowance for free-surface effects has modified the solutions.

<u>The results</u>

```
PATCH(   6,CONTUR,    1,  15,    1,  10,    1,    1,    1,    1)
COVAL(   6,P1   , 0.000E+00, 5.000E+01)
CONTOUR PLOT OF P1
  Z-PLANE: X HORIZONTAL, Y VERTICAL
MINIMUM =-1.945E-01 MAXIMUM = 2.617E-01
PHOENICS-84 RECTANGULAR CONTOUR PLOT
N,S,E,W... DENOTE: NORTH, SOUTH, EAST ..

 0 NNNNNNNNNNNNNNNNNNNNNNNNNNNNNNNNNNNNNNNNNNNNNNNNNNNNNNNNNNNNNN
 1 W  00000      888       6666666666          88888         0000000000    E
 2 W  000000     888       66666666666         88888         000000000     E
 3 W   00000     888      666666666666         88888         000000000     E
 4 W    00000    888     66666666666666        88888         000000000     E
 5 W     0000    88      66666666666666        88888         000000000     E
 6 W      000    88     666666666666666        88888        0000000000     E
 7 W       000 88   666666        666666        888         000000000      E
 8 W   222   00 8   666            66666        888        0000000000      E
 9 W  22222   0 8 666                6666      888        000000000       E
10 W  2    22 0 8 6                  666      88        000000000        E
11 W 22 44 22    6  44444444444        66    888       000000000         E
12 W 2 4  4            4444444444      66    888      00000000           E
13 W 2  66 4    6  222      444444444  6 . 88      0000000             E
14 W 246   6 4284 8 0 00        444444 6   88    00000                E
15 W      8 6 2 2                     4 6   8   0000                   E
16 W   6800 884826                     4   88 000                      E
17 W 4   244 6    6                    4   8   0                       E
18 W 4   4   24260                     4   8 00      22                E
19 W 48   8   64- -- - ---- - ---- - ---- -46 8 0   222222222          E
20   SSSSSSSSSSSSSSSSSSSSSSSSSSSSSSSSSSSSSSSSSSSSSSSSSSSSSSSSSSSSS
```

Fig.1 Floating vessel, rigid upper water surface.

Fig. 1 represents the pressure distributions for case 1. The lower "south" edge of the plot is the symmetry axis; the presribed inflow of water is from the left ("west") and the outflow is to the right ("east"), where the pressure is held uniform. The upper ("north") edge of the diagram represents the canal wall. Since the north-south dimension of (one half of) the integration domain is 5.0m whereas its east˜west dimension is 15.0m, it is evident that the contour plot is artificially compressed in the latter direction.

The legend at the top of the diagram indicates that the maximum pressure in the field is 0.2617 and the minimum is 0.1945 above the undisturbed water pressure. The units are those of ρu^2_∞, where u_∞ stands for the approach velocity, so that a maximum of 0.5 would be expected to prevail; the coarseness of the grid, and the fact that no grid point is located at the stagnation point, accounts for the lower value.

The contour-plotting procedure was set to divide the whole range of pressure into fifty intervals, and to print only the second digit of the interval number, if even. This is why bands of 0's, 2's, 4's, 6's, and 8's appear on the diagram. Interpretation of the diagram, assisted by inspection of tabulations of the pressure fields, reveals that the maximum pressure is found just ahead of the object, while the lowest pressure appears at its flank. At the stern of the object there is a low-pressure wake, in which recirculation is present. Along the north boundary the pressure falls and then rises in a fairly symmetrical manner.

There is no need to attempt to extract more information from Fig. 1, which has been provided mainly in order to permit comparison with Fig. 2, which represents the pressure distribution in the upper layer when free-surface effects are allowed for.

```
PATCH(   6,CONTUR,   1,  15,   1,  10,   1,   1,   1,   1)
COVAL(   6,P1   , 0.000E+00, 5.000E+01)
CONTOUR PLOT OF P1
  Z-PLANE; X HORIZONTAL, Y VERTICAL
MINIMUM =-1.401E-01 MAXIMUM = 3.198E-01
PHOENICS-84 RECTANGULAR CONTOUR PLOT
N,S,E,W... DENOTE: NORTH, SOUTH, EAST ..

 0 NNNNNNNNNNNNNNNNNNNNNNNNNNNNNNNNNNNNNNNNNNNNNNNNNNNNNNNNNNNNNNNN
 1 W            44  22  00  88    66666   88   00  22    4444    E
 2 W            44  22  0  888   666666   88   00 222   44444    E
 3 W     6666   44  22  0  88    666666   88  00  222   44444    E
 4 W    666666  44  22 00  88   6666666   88  00  222   4444     E
 5 W    666666   4  22 00 88   66666666   88  00  22    4444     E
 6 W   66666666  4  2  0  88  666666666   88  00  22    4444     E
 7 W   6666  66  4  2 00 88  6666  66666  88  00 222   44444     E
 8 W  66       66 4 22 0 88  666     6666  88 00  22   44444     E
 9 W  66   88  6 4 2 0  8  66        666  8  00  22   44444     E
10 W  6 888888 6 4 2 0 8 666          66  8  0  22   44444      E
11 W 66 88      6 42 0 8 66   444444  66 88 00  22  444444      E
12 W 6 8  000 004 20 8 66  444444444  66 88 0  22   444444      E
13 W 6 800   08642  866  4444     44  66 8  0 22   4444444      E
14 W 68   22  86 0 86 444          44  6 8 0  2   4444444       E
15 W  802  442                    2  4   8 00 2   44444444      E
16 W     4  8 0 0                22 4 6 8 0 22  444444444      E
17 W  024 0 6                    2 2  6 80 22  4444444444      E
18 W  0 604   4                   02  6  022  44444444444      E
19 W 8  6 60   - 0 - - - - - - 2 - -02 46  022  44444444444     E
20  SSSSSSSSSSSSSSSSSSSSSSSSSSSSSSSSSSSSSSSSSSSSSSSSSSSSSSSSSSSSSSS

20   2 4 6 8 0 2 4 6 8 0 2 4 6 8 0 2 4 6 8 0 2 4 6 8 0 2 4 6 8 0
```

Fig.2 Floating vessel; pseudo-compressible water surface.

Figure 2 is a plot of exactly the same kind; but the positions of the contour lines are different from before, as indeed are the maximum and minimum pressures.

What is evident is that there is a greater tendency for the contour bands to slope up from left to right, which is reminiscent of the way in which waves do actually appear in the vicinity of a surface ship.

```
PATCH(   6,CONTUR,   1,  15,   1,  10,   1,   1,   1,   1)
COVAL(   6,P1   , 0.000E+00, 5.000E+01)
CONTOUR PLOT OF P1
  Z-PLANE; X HORIZONTAL, Y VERTICAL
MINIMUM =-1.862E-01 MAXIMUM = 1.090E-01
PHOENICS-84 RECTANGULAR CONTOUR PLOT
N,S,E,W... DENOTE: NORTH, SOUTH, EAST .

 0  NNNNNNNNNNNNNNNNNNNNNNNNNNNNNNNNNNNNNNNNNNNNNNNNNNNNNNNNNNNN
 1 W   0000   88  666    4444444      666   8888     000000    E
 2 W   0000   88  666   44444444      666   8888     000000    E
 3 W    0000  88  66    444444444     666   8888     000000    E
 4 W     000  88  66   44444444444    666   888     0000000    E
 5 W      000 88 666  444444444444    666   888     0000000    E
 6 W   2  00 88 66  44444    44444     66  8888     0000000    E
 7 W  2222  0  8 66 4444       4444    66  8888    00000000    E
 8 W  22 22  0 8 6  44           444   66  888     0000000     E
 9 W 22    22 0 86  44    2222     44  66  888    00000000     E
10 W 2 444  2   6 4  22222222222   44 66  888     0000000      E
11 W 2 4  4      4  22222  2222222 44 66  888    00000000      E
12 W 24  6  42  4222           222 4 66  88    00000000        E
13 W   688 6   6   000000000     2 4 66 88   000000000         E
14 W    0 8  406 - --- - -000000000 2  6  8   00000000         E
15 W 46   00  0 |66  888        | 02  6 88  000000             E
16 W 4 0  22  0 | 44 6  888     88  6 8 0000                   E
17 W  8    44  24| 02   6 8888888 | 80   88 000                E
18 W            |     4 6    8  66 02  8 0                     E
19 W    888882-0--80-44-6---  - -66- 2  8 0                    E
20  SSSSSSSSSSSSSSSSSSSSSSSSSSSSSSSSSSSSSSSSSSSSSSSSSSSSSSSSSSSS

20   2 4 6 8 0 2 4 6 8 0 2 4 6 8 0 2 4 6 8 0 2 4 6 8 0 2 4 6 8 0
```

Fig.3 Submerged vessel, rigid upper water surface.

```
PATCH(   6,CONTUR,    1,  15,   1,  10,   1,   1,   1,   1)
COVAL(   6,P1   , 0.000E+00, 5.000E+01)
CONTOUR PLOT OF P1
  Z-PLANE; X HORIZONTAL, Y VERTICAL
MINIMUM =-1.528E-01 MAXIMUM = 1.402E-01
PHOENICS-84 RECTANGULAR CONTOUR PLOT
N,S,E,W... DENOTE: NORTH, SOUTH, EAST ..

 0 NNNNNNNNNNNNNNNNNNNNNNNNNNNNNNNNNNNNNNNNNNNNNNNNNNNNNNNNNNNNNNN
 1 W6666       666 44 2 00 8 6  44          4  6  8 0  2  44      6E
 2 W666         66 44 2 0 88 6  44    22    4  6  8 0  2 444      6E
 3 W666         66 44 2 0 88 6  44   222    4  6  8 0 22 444      6E
 4 W666    88   66  4 2 0 8 66 44  22222   44 6 88 0 22 44       66E
 5 W666  88888  66 4 2 0 8 6  4  2222222   44 6 8 00 2  44       66E
 6 W66  8888888 66 4 2 0   6 44  2222222   4  6 8 0  2 444      666E
 7 W66  888    8  6 4  0 8 6  4  22     2  4  6 8 0 22 44     66666E
 8 W66 88      88 6 42 0 8 6 4 222       22 4 6 88 0 2  4     66666E
 9 W66 8  0000 8 6  2 08 6 4 22  0000    2 4 6 8 0 2  44     666666E
10 W6 8 000    08 6  20 8  4  2  0000000 2 4 6 8 0 2 44      666666E
11 W6 8 0  2220 864  08 644 2 00       00 2 4 68 0 2  44    6666666E
12 W6 80 2  4   8642  86 4 2 00  888     2 4  8 0 2 44     66666666E
13 W6   2 44 420864208  4 2 0  88  888 02 46  0 2 44     666666666E
14 W68   4 6 6  86- -  64 2 0 88  66  8 02  68  2  4    6666666666E
15 W6802 6 8 8  8   86  2 0 8  6 | 66 8024  8022 4      6666666666E
16 W68  68 020628  08 4  0 8 66 4| 46 802468  2 44      6666666666E
17 W68 4   2428406  864  0 8 6 4 22  680   80   44      6666666666E
18 W6 2 8   6 0 | 4 8642  8 6 4 20  468  68  2 44     66666666666E
19 W6 26    86-6--4-8642--8-64 2-0- 6  4680 244     666666666666E
20  SSSSSSSSSSSSSSSSSSSSSSSSSSSSSSSSSSSSSSSSSSSSSSSSSSSSSSSSSSSSS

20   2 4 6 8 0 2 4 6 8 0 2 4 6 8 0 2 4 6 8 0 2 4 6 8 0 2 4 6 8 0
*****************************************************************
```

Fig.4 Submerged vessel; pseudo-compressible water surface.

Figures 3 and 4 permit a comparison of the same kind to be made for the cases in which the body is submerged. When the rigid-surface presumption is made, the pressure contours exhibit a tendency to be symmetrical about a vertical line; whereas the pressure contours for the free-surface presumption exhibit a pronounced sloping tendency.

```
PATCH(   6,CONTUR,   1,  15,   1,  10,   1,   1,   1,   1)
COVAL(   6,P1   , 0.000E+00, 5.000E+01)
CONTOUR PLOT OF P1
  Z-PLANE; X HORIZONTAL, Y VERTICAL
MINIMUM =-1.387E-01 MAXIMUM = 4.473E-01
PHOENICS-84 RECTANGULAR CONTOUR PLOT
N,S,E,W... DENOTE: NORTH, SOUTH, EAST ..

  0 NNNNNNNNNNNNNNNNNNNNNNNNNNNNNNNNNNNNNNNNNNNNNNNNNNNNNNNNNNNNNNN
  1 W 2 4 6 80 2 44        44 2  0 8 6 44 2  0  8  66     444      80 E
  2 W 2 4 6 80 2 44       44  2 0  8 6 4  2 00 88  66     44       80 E
  3 W 2 4 6  0 2 4        44 22 0 8  6 4 22 00 88  66     44     6 80 E
  4 W 2 4 68 0 2 4      44  2 00 8 6  4 2  0  88  66     44     6  0 E
  5 W 2 4  8  2  4    44  2  0 8  6 4  2  0  8  666            6  0 E
  6 W 24 6 80 2 44 4444  22 0  8 6  4 2  0  88  666           66  0 E
  7 W  4 68 02  44 444  22 0  8 66 4  2 00  88  666          66 8 0 E
  8 W  4  80   444 44  2  0  8 66 4  2  0  88  6666         666 8 0 E
  9 W  46  02 444 44  2  0 88 66 4  2  0  88   6666     6666  8 0 E
 10 W   680  4   44  2 00 8  6  4  2  0  88   66666666666   8 00 E
 11 W      24 6  44 2 0 88 66  4  2  0  88   6666666666   888 0  E
 12 W  6802  666 4 2 0 8  6  44 22 00 88   666666666    888  00  E
 13 W 46   46  64  0 8 66  44  22 0  8   66666666    8888   000  E
 14 W 4 024  886  00 0    44    00 0 88 6666666666   888    0000   E
 15 W 4  4  808 2                   666    6666  888    000000     E
 16 W  82   0206                     4444   666 888   000000       E
 17 W    6  6 60                    2  44 66  8   00000            E
 18 W      260 2                    0 2 4 6 88  000                E
 19 W 6048  26-46 . ---- . ---- . --.2 2 4.66 8  000               E
 20  SSSSSSSSSSSSSSSSSSSSSSSSSSSSSSSSSSSSSSSSSSSSSSSSSSSSSSSSSSSSSS

 20   2 4 6 8 0 2 4 6 8 0 2 4 6 8 0 2 4 6 8 0 2 4 6 8 0 2 4 6 8 0
```

Fig.5 Floating vessel in shallow water.

Finally, Fig. 5 represents the result of the calculation in which the shallow-water presumption is made. Now the pressures can be regarded as being proportional to the square of the water-surface-height differences induced by the motion of the object.

If lines are drawn through chains of 0's, 2's, 4's, etc, their pronounced slope will again be perceived. Their pattern is reminiscent of the well-known compression waves which surround a body moving at supersonic speed through a gas.

Discussion

Limitations of space and time have rendered it impossible to do justice, in the present article, to even the unambitious series of calculations which have been

performed; but it is in the nature of numerical simulations of fluid flow that documentary reporting is much less effective than the making available of the computer program and its unputs.
In the present case, the computer code PHOENICS is available on a royalty-free basis to academic research institutes; and the author will be willing to send a copy of the input file defining the above calculations to anyone who requests it. There is much work still to be done before the method is validated.

As to the significance of the results which have been shown, the author believes it to be as follows:-

(a) The calculations have shown that it is as easy, in Navier-Stokes solution procedures, to take account of free-surface effects as to neglect them.
(b) The calculations exhibit free-surface effects which are qualitatively of the expected kind.
(c) Finer-grid studies will be needed before it becomes worthwhile to make detailed comparisons between the predictions and experimental data.
(d) Where such comparisons are made, it may well be found that further refinements will be needed, involving additional use of the blockage-factor notion, in order to procure precise agreement.
(e) Which of the three methods described above (viz. GALA, blockage, and pseudo-compressibility) will be found the most satisfactory is still an open question.

References

1. Lindgren, H.; (Chairman Org. Comm.) Second International Symposium on Ship Viscous Resistance, Goteborg, Sweden, 1985.

2. Maxwell, T.T.; Numerical modelling of free-surface flows. 1977.

3 Pericleious, K.A.; Unpublished work at CHAM LTD. 1985.

4. Rosten, H.I. & Spalding, D.B.; Beginners guide to PHOENICS-84". CHAM Ltd Technical Report, no. 99. 1985.

5. Spalding, D.B.; A Method for computing steady and unsteady flows possessing discontinuities of density. CHAM Report 910/2. 1974.

6. Yeung, R.W.; Numerical methods in free-surface flows, in Ann. Rev. Fluid Mech. vol.14 395-442. 1982.

Large-Scale CYBER-205 Simulation of Vortex Flowfields Around Submarines

Arthur Rizzi
FFA The Aeronautical Research Institute of Sweden,
S-161 11 BROMMA, Sweden
and
Charles J. Purcell
ETA Systems, Inc., St. Paul, MN 55112, U.S.A.

Abstract

There is great practical incentive to simulate by computer the flow around a submarine. Potential models currently in use are limited in their treatment of separation. According to Maskell's classification the flow leaves the surface in either a bubble-type (closed streamline) separation or a free-shear-layer (open streamline) separation. We hypothesize that open separation is an inviscid but rotational phenomenon that is correctly described by the incompressible Euler equations. The large-scale numerical simulation using over 600,000 computational cells which is presented here for flow around a 9.5:1 ellipsoidal body supports the conjecture.

Introduction

A submarine traveling under water shares many of the same characteristics as an airplane in flight. Both are surrounded by homogeneous media, both alter their vertical position by creating vorticity on their plane surfaces, and both must be stable while sustaining the forces and moments acting on them (see Fig. 1). And just like his airplane counterpart, the submarine designer wants additional insight and information from numerical simulations to help him with his problems. For most flows with marine applications the Reynolds number is extremely high, typically around a billion. Boundary layers therefore are extremely thin. Neglecting this viscous layer, the standard inviscid simulation model for such flows is the equation for irrotational flow together with some means for incorporating the vorticity generated by the ship, either by the use of discrete vortices or by fitting a vortex sheet to the potential-flow soluton (see Lugt[1] for a review of these models). A novel alternative to this approach is to adopt the

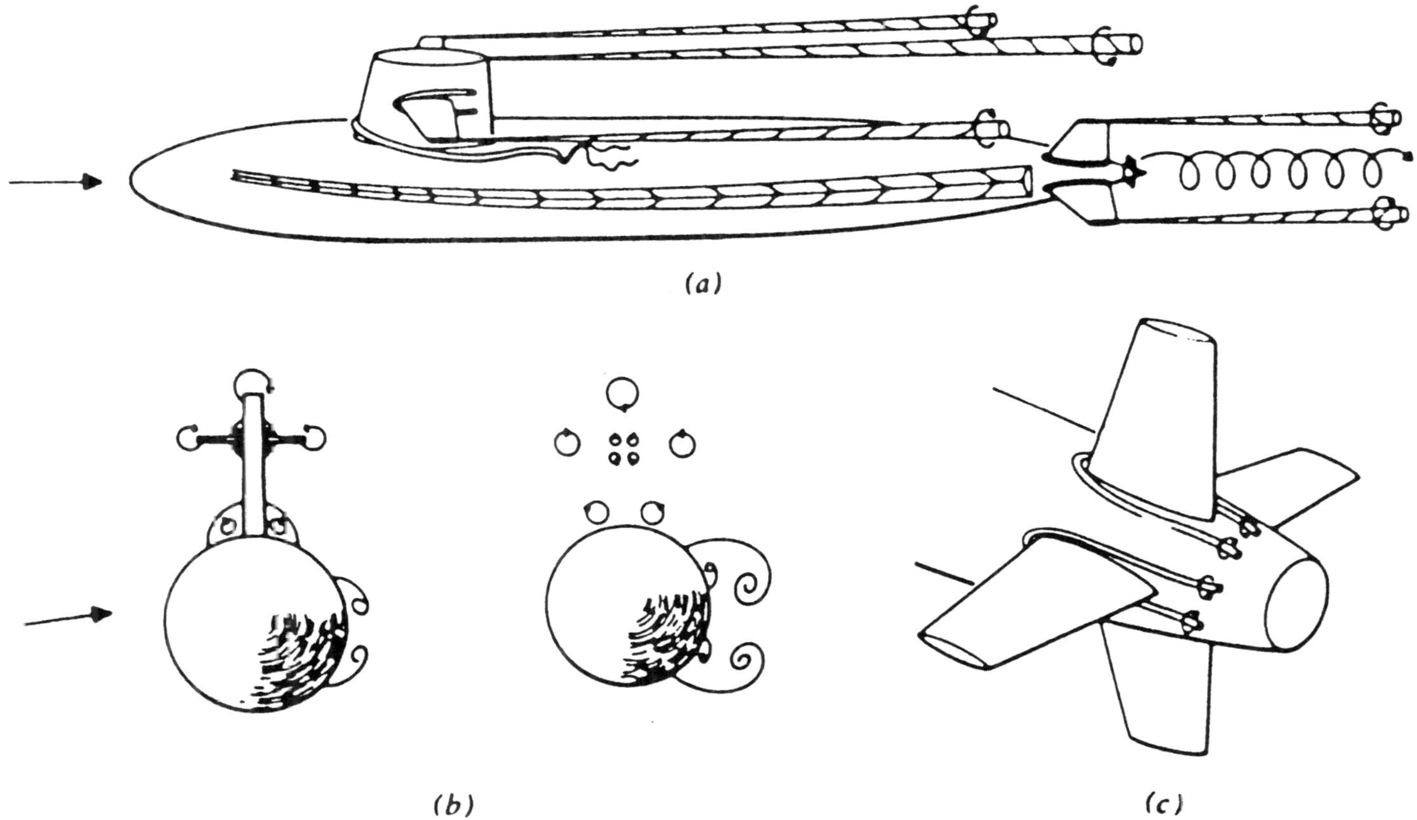

Fig. 1a. The major vortex configurations around a submarine et a small angleo of attack in a slight turn. b) Two cross sections of the hull at the sail and behind. In the latter case the local angle of attack (drift angle) is larger. c) Necklace vortices around the control surfaces at the stern (from H.J.Lugt (1))

Euler equations of incompressible flow as the model. It offers the advantage that vorticity and vortex sheets are admissible parts of its continuous and discontinuous solutions, respectively. But how the vorticity and vortex sheets get created initially is a matter of controversy, which usually centers around the claim that vorticity in an incompressible fluid can only be generated by viscous forces acting in the boundary layer. We take the contrary view point here. Unlike the Laplace equation, the Euler equations admit solutions on the surface of the body with separation lines that are the start of vortex sheets. Once started, a vortex sheet coils up under its own influence to form a vortex. Maskell originally classified this type of separation line as regular and termed it the free-vortex layer since all the streamlines are open. The second type, bubble separation, he called singular and its streamlines are closed.

Recently we have developed a pseudo-time hyperbolic method to solve the incompressible Euler equations[3]. That work established that open separation with a vortex sheet is captured in the numerical simulation of flow past a body with a salient edge. The intent here is to explore whether the same phenomena is observed when the body is a smooth submarine-like shape. The advent of supercomputers with very large real memory just now opens up to us the possibility to explore numerically controversial issues like this one. A realistic test of course requires a three dimensional discrete model of significant size. The recent construction of a 16-million-word memory for the CYBER 205 (or the NASA-VPS32) has allowed tests of one-million-grid-point models on a practical basis within reasonable elapsed times. Prior to this it was hoped that this type of computation could be performed with disk I/O but experience showed that heavy time penalties were encountered due to the demands of the flow-solving algorithm. Therefore good virtual-memory-management techniques within a sufficient working set of real memory seems to be the only effective way to carry out such large-scale simulations. The other crucial requisite is a large band-width communication channel between the memory and the arithmetic unit.

Open or Closed Separation

Consider the various types of separation phenomena illustrated in Fig. 2. A rounded slender body meets an oncoming stream at an angle of incidence that varies from low to high. We conjecture that the Euler equations describes open-streamline separation, but not closed (or bubble-type) because then the problem is not well-defined, since in the absence of diffussive processes, unique conditions cannot be set on these closed streamlines. We test this hypothesis by searching for numerical evidence of such open-streamline separation. This requires good numerical resolution, and subsequent detailed examination, of the flow near the body surface.

DEVELOPMENT OF MATHEMATICAL MODEL

Continuum Equations

The complete model to describe the fluid mechanics of this vortex flow is the incompressible Navier-Stokes equations:

$$\operatorname{div} \underset{\sim}{V} = 0$$

$$\frac{\partial}{\partial t} \underset{\sim}{V} + \operatorname{div} \underset{\sim\sim}{VV} + \operatorname{grad} \frac{p}{\rho_o} = \Gamma \qquad (1)$$

$$\frac{\partial}{\partial t} e + \operatorname{div} (e+p)\underset{\sim}{V} = T$$

where ρ_o is the constant density, p the pressure, e the energy, and $\underset{\sim}{V}$ the velocity of the fluid. The term Γ is the shear stresses acting on the fluid and T is heat conduction and viscous dissipation terms. If the Reynolds number of the flow is large, say over one million, the terms Γ and T have an effect only in very thin layers of the flow near the surface of the body and across the shear layers that separate from it. Because these free shear layers are thin, Γ and T do not have to be accurately represented, but can be modelled instead by simpler nonphysical expressions. The boundary layer, however, cannot be represented in such a model. Equations (1) then are

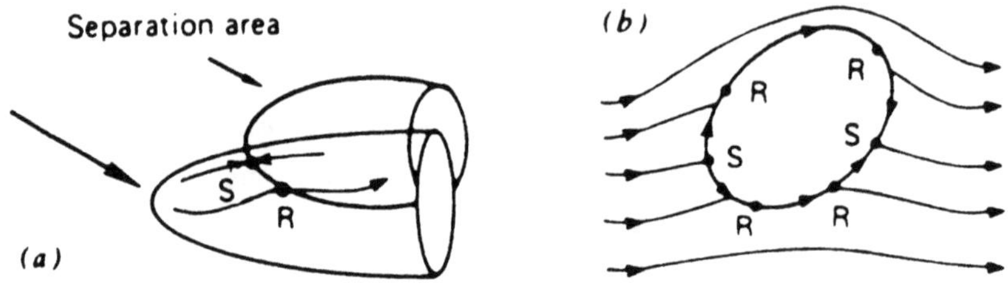

(S) Singular and (R) regular separation points on a body surface. They form a "closed" separation line: (a) in perspective and (b) seen from above.

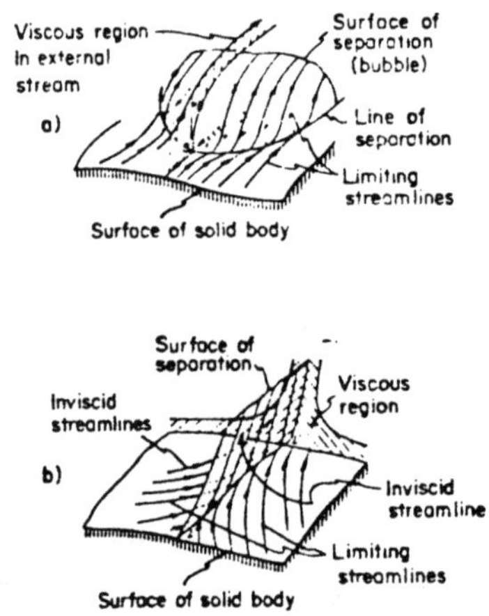

Maskell's separation concepts. a) Bubble, b) Free vortex layer.

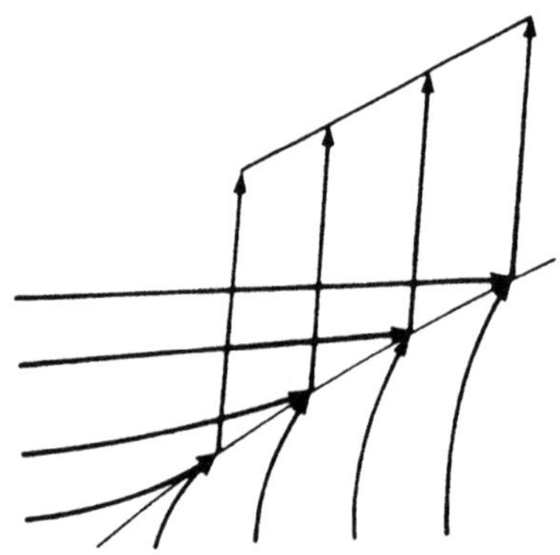

"Open" separation line.

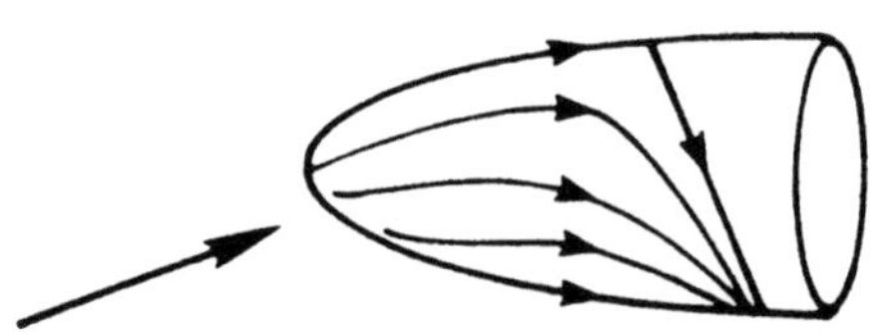

The separation line is a streamline.

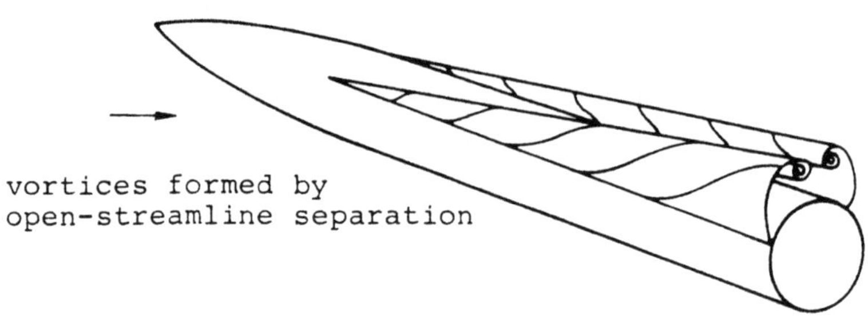

Fig. 2. Open and closed types of separation phenomena (from Lugt (1))

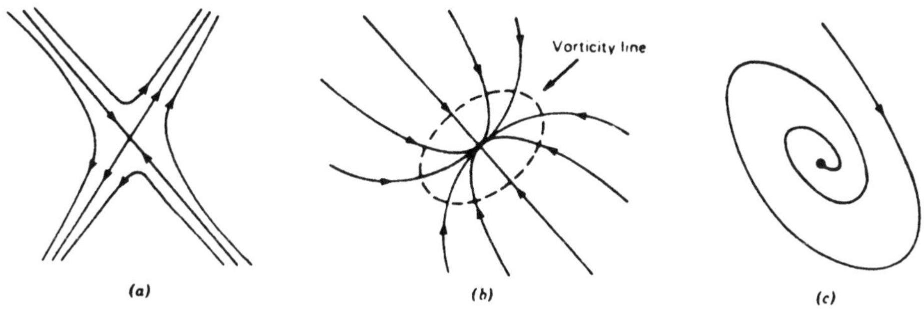

Singular separation points: (*a*) nodal point, (*b*) saddle point, and (*c*) spiral point.

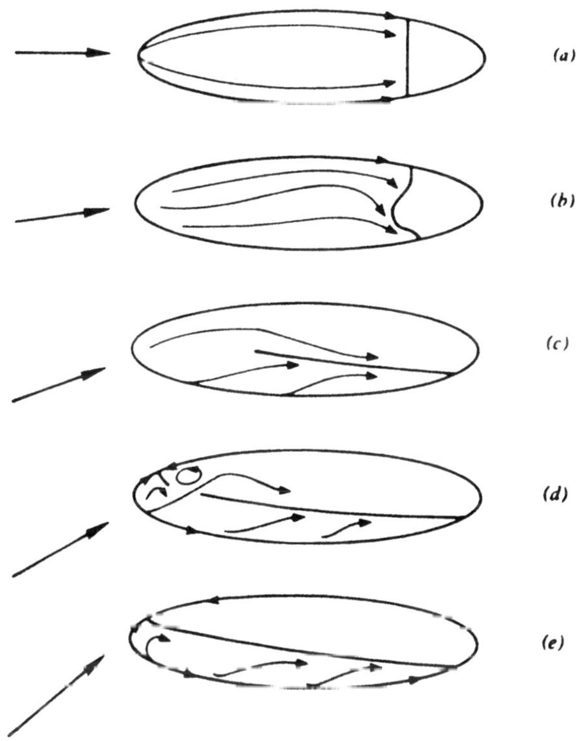

Flow around a prolate spheroid at various angles of attack. In cases (*a*), (*b*), and (*e*) the separation lines are closed; in (*c*) and (*d*) they are open

Figure 2.

called the Euler equations with an artificial viscosity model. The flow is inviscid except across thin shear layers that are admitted in the solution where irreversible processes can take place.

Mesh Generation

The first step in carrying out a numerical solution of Eqs.(1) is to discretize the flowfield by creating a mesh that spans the entire region. We use Eriksson's method of transfinite interpolation[4] to construct a two-pole boundary-conforming O-O grid around a 9.5 to 1 ellipsoidal body. This mesh has one polar singular line attached to its nose and another polar singular line at its tail (see Fig. 3).

Numerical Solution Procedure

The pseudo-time dependent method to obtain steady flow solves the incompressible Euler equations in finite-volume form

$$\frac{\partial}{\partial t}\iiint q \; dvol + \iint \underset{\sim}{H}(q)\cdot\underset{\sim}{n} \; dS = \begin{matrix}\text{artificial}\\ \text{viscosity model}\end{matrix} \qquad (2)$$

where q is the vector with elements [p, u, v, w] for pressure and Cartesian components of velocity with reference to the fixed coordinate system x,y,z. The flux quantity $\underset{\sim}{H}(q)\cdot\underset{\sim}{n}=$ $[(c^2,u,v,w)\underset{\sim}{V}+(0,\underset{\sim}{e}_x,\underset{\sim}{e}_y,\underset{\sim}{e}_z)p]\cdot\underset{\sim}{n}$ represents the net flux of q transported across, plus the pressure p acting on, the surface S surrounding the volume of fluid. The constant c^2 is an arbitrary parameter that can be chosen to accelerate convergence. The mesh segments the flowfield into very many small six-sided cells, in each of which the integral balance (2) must hold. The finite volume method then discretizes (2) by assuming that q is a cell-averaged quantity located in the center of the cell, and the flux term $\underset{\sim}{H}(q)\cdot\underset{\sim}{n}$ is defined only at the cell faces by averaging the values on each side. With these definitions and calling the cell surfaces in the three coordinate directions of the mesh $\underset{\sim}{S}_I$, $\underset{\sim}{S}_J$, and $\underset{\sim}{S}_k$, we obtain the finite volume form for cell ijk

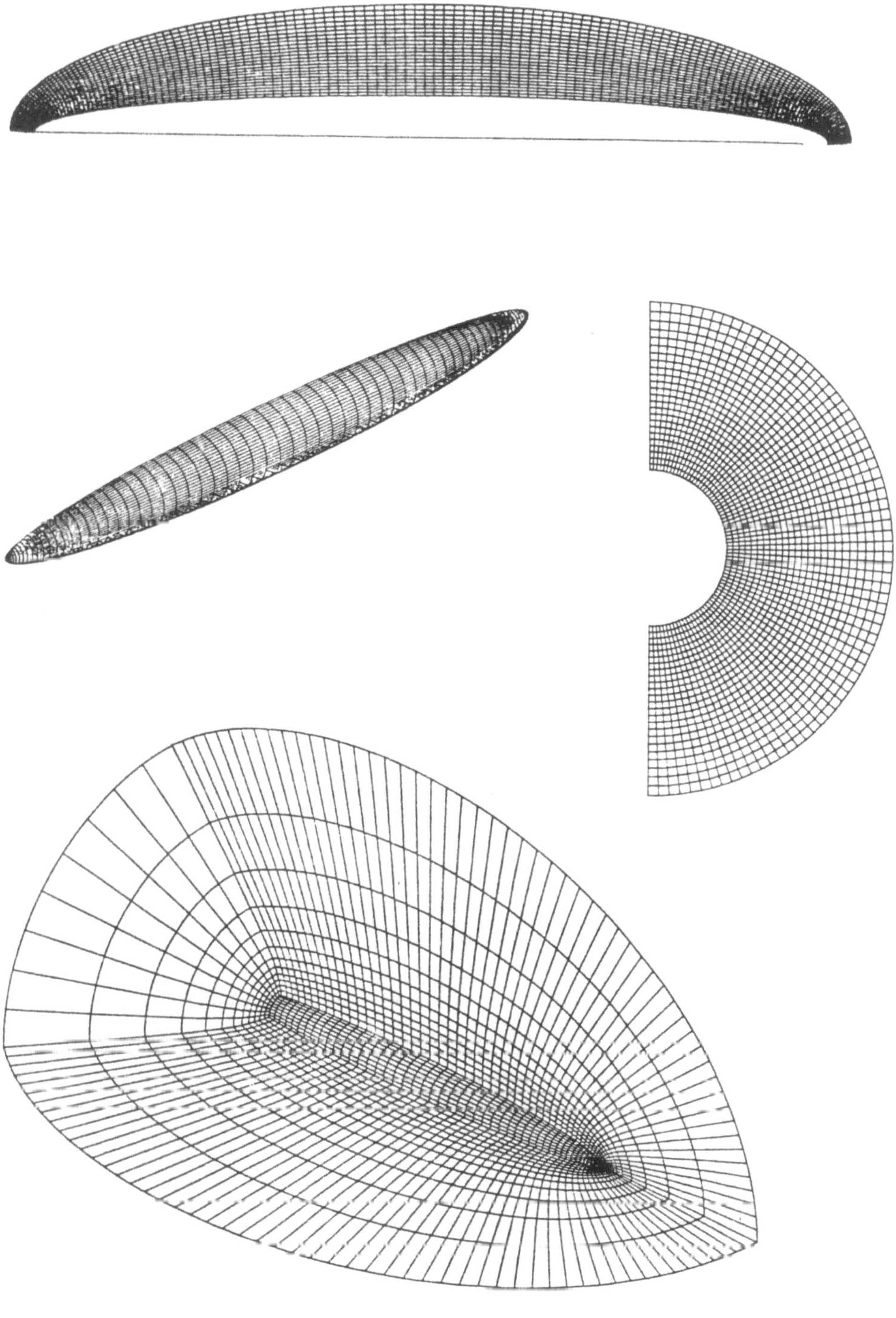

Fig. 3. Partial views of the two-pole O-O type mesh around a submarine-like shape used in the computations. Contains 73 x 61 x 145 points

$$\frac{\partial}{\partial t} q_{ijk} + [\delta_I(\underset{\sim}{H}\cdot\underset{\sim}{S}_I) + \delta_J(\underset{\sim}{H}\cdot\underset{\sim}{S}_J) + \delta_k(\underset{\sim}{H}\cdot\underset{\sim}{S}_k)]_{ijk}\cdot\underset{\sim}{n} = \begin{array}{c}\text{aritificial}\\ \text{viscosity model}\end{array} \quad (3)$$

where $\delta_I(\underset{\sim}{H}\cdot\underset{\sim}{S}_I) = (\underset{\sim}{H}\cdot\underset{\sim}{S}_I)_{i+1/2} - (H\cdot S_I)_{i-1/2}$ is the difference operator. To this the boundary conditions for the particular application must be specified. They occur at the six bounding faces of the computational domain. When included we can write Eq.(3)

$$\frac{\partial}{\partial t} q_{ijk} + FD = \begin{array}{c}\text{artificial viscosity model}\\ +\\ \text{boundary conditions}\end{array} \quad (4)$$

where FD represents all of the flux differences. A more detailed description of the method is given in Ref. 3. With the spatial indices suppressed we integrate this last equation with the two-level three-stage scheme in parameter $\theta=1/2$ or 1

$$\begin{aligned} q_o &:= q \\ q &:= q_o + \Delta t\ FD(q_o) \\ q &:= q_o + \Delta t[(1-\theta)FD(q_o) + \theta\ FD(q)] \\ q &:= q_o + \Delta t[(1-\theta)FD(q_o) + \theta\ FD(q)] \end{aligned} \quad (5)$$

that steps the solution forward in time until a steady state is reached.

Data Structure and Methodology for Vector Processing

The nature of the vector instruction in the CYBER 205 emphasizes rapid operations upon contiguous cells in memory. A three dimensional structure suitable for vector processing is correctly visualized as consisting of a collection of adjacent pencils of memory cells with suitable boundary conditions This leads to the natural grouping for the formation of differences in J by means of differencing adjacent pencils. The natural differencing in the K-direction thus becomes the difference of a plane of adjacent pencils. Differences in the I direction are formed by use of single elements offset one against the next within the pencil. These relationships then allow for the complete vectorization of all of the internal difference expressions. In addition other relationships have to be defined on the boundary. Computation of these external, i.e. boundary, conditions interrupt the flow of vectorization.

Our particular choices in conjunction with the data-base design that overcomes this interruption of vector processing lead to the following key features of the procedure: 1) separate storage arrays are assigned for the dependent variables q, flux component F, and flux differences FD, 2) one extra unit is dimensioned for each computational direction to hold the boundary conditions, and 3) flux differences are taken throughout the entire field by off-setting the starting location of the flux vector F (see Ref. 5). In this way all of the work in updating interior points is exclusively vector operations without any data motion, accomplished at the expense of just a single scalar loop for the boundary conditions on one computational direction and a doubly-nested loop for another, while the boundary conditions for the third direction vectorize completely. The generalized mapping from data to computer memory is managed via common block declarations within FORTRAN. The common blocks are grouped into large pages by the loader. The large pages are stored on the disk system across a series of independent temporary files, each of which is accessed and mapped into the executable space.

SIMULATED FLOWFIELD WITH OPEN SEPARATION

Incompressible flow past the 9.5 to 1 ellipsoid at 20 deg. incidence is simulated using a grid of 72×60×144 cells, and the results for pressure, velocity, and vorticity are displayed in a variety of views in Fig. 4. Details in the character of the separation show up most lucidly in the velocity field and can be seen best by probes of velocity-vector components drawn near the body surface. Contours of isovorticity magnitude complement these views and also indicate the presence of the vortex sheets. Figure 4 gives strong numerical evidence for the likelihood that the Euler equations do describe open-type or free-shear-layer separation. This model then becomes very interesting for use in the practical design of a submarine, e.g. predicting forces and moments on the ship when it is in a constant-radius turn, as well as quantitatively analyzing the structure of its vortex wake.

REFERENCES

1. Lugt, H.J.: Numerical Modelling of Vortex Flows in Ship Hydrodynamics, A Review, Proc. Third Int'l Conf. on Numerical Ship Hydrodynamics, Paris, 1981.

2. Maskell, E.C.: Flow Separation in Three Dimensions, RAE Report No. AERO. 2565, London, 1955.

3. Rizzi, A.W. and Eriksson, L.E.: Computation of Inviscid Incompressible Flow with Rotation, Journal Fluid Mechanics, Vol. 153, April 1985, pp. 275-312.

4. Eriksson, L.E.: Generation of Boundary-Conforming Grids around Wing-Body Configurations using Transfinite Interpolation, AIAA J., Vol. 20, Oct 1982, pp. 1313-1320.

5. Rizzi, A.: Vector Coding the Finite-Volume Procedure for the CYBER 205, in Lecture Series Notes 1983-04, von Karman Institute, Brussels, 1983.

Session 6 Navigation Methodology

Simulation of a Digital Filter/Estimator for the Navigation of Large Ships in Confined Waters

M J DOVE, R S BURNS, C T STOCKEL, T H BOUNCER

Ship Dynamics and Control Research Group, Plymouth Polytechnic, and D J MacKinnon, Rutherford-Appleton Laboratories, SERC.

Summary

Aeronautical and marine casualty statistics indicate that the human being, when under stress or at times of peak load, can be a poor co-ordinator of the information available to him, particularly when that information is from a number of different sources, as is often the case in modern ships. Integration and co-ordination of information and its useful application in a closed loop feedback system can reduce the probability of accident as has already been demonstrated in the case of automatic landing systems for aircraft.

This paper describes the development of a digital filter/estimator for use in conjunction with an optimal controller in the navigation of large ships in the approaches to a port.

Introduction

It is beyond question that the overall standard of navigation at sea is very high indeed, and the probability of completing a voyage successfully must be very close to unity. However, (1), a brief summary of marine traffic accidents shows that the majority occur within congested waters, particularly within port limits. Congestion, coupled with the increased size and complexity of operation, has focussed attention on the control of pilotage and berthing, for, not only must the safety and cost factors be considered, but also the environmental aspects of, say, the spillage of large quantities of crude oil at, or near, the approaches to a port.

A discrete, time-varying non-linear model, (2), has been developed. This is based upon eight system states, namely forward and lateral position and velocity, heading, yaw-rate, rudder angle and engine speed. The model has two deterministic inputs - demanded rudder and engine speed plus four stochastic disturbance inputs in the form of wind and current vectors. The measurements of the state vector, contaminated with random noise, are passed through an optimal, time-varying filter.

The best estimate of each state variable is used by an adaptive optimal controller to compute those inputs (demanded rudder and engine speed) which minimise a given performance criterion. The dynamics of both the filter and controller are updated frequently by a system identification algorithm that can be either based upon apriori knowledge of the hydrodynamic coefficients of the vessel, or by on-line measurements of the state variables.

An outline of the system is given in Figure 1.

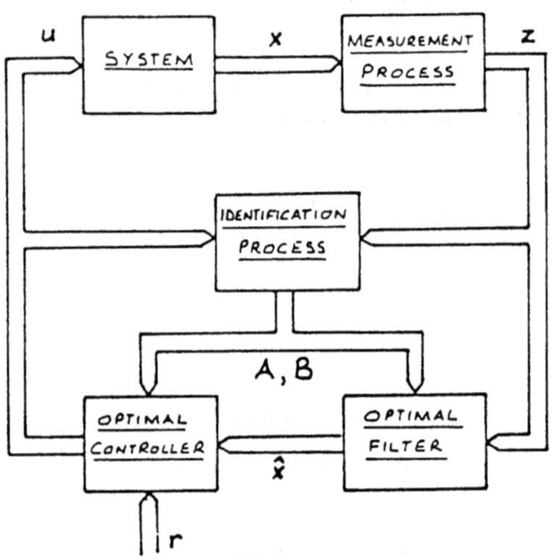

Figure 1.
Proposed automatic guidance system

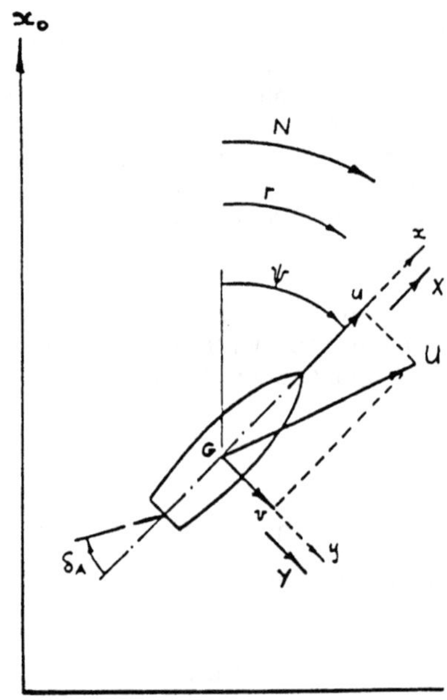

Figure 2.
Co-ordinate systems

Co-ordinate Systems and Sign Conventions

Two right-handed co-ordinate systems are used, the first with respect to the ship (x,y) and the second with respect to the sea bed $(x_0, \dot{y}_0)$. These are shown in Figure 2 with the positive directions as indicated. The origin of the ship co-ordinate system is assumed to be at the ship's centre of gravity. The axes of the earth co-ordinate system conform with standard navigational practice, ie, the x_0 axis corresponds to the direction of True North. The positive directions are as given in Figure 2.

In modelling disturbance inputs such as wind, waves, current and depth of water it is assumed that in the approaches to a port that wave excitation can be ignored, accelerations of current and wind are small enough to be neglected, and the depth of water is such that the mathematical model is not affected.

Integration of Navigational Data

Single system deficiencies have led to the development of integrated systems for world wide use, (3). In shore, particularly in the approaches to a port, and in the development of off-shore energy resources, there is a much greater need for accurate navigational data, giving a further impetus to the development of integrated systems. If it is assumed that the systematic errors can be allowed for then the requirement of an integrated system is to minimise in some way the random errors. A Gaussian distribution gives the best general fit for the spread of random errors and this implies a definition of these errors in terms of standard deviations or root mean square errors. As variance is the square of standard deviation the problem can be stated in terms of minimising the variance, which has led to the use of minimum variance or Kalman-Bucy filters. These have been developed extensively for aerospace, and latterly marine, (4), (5), navigation since the publication of the original work by Kalman and Bucy, (6).

Measurement and Filtering

The measured state $\underline{z}(k + 1)$ is considered to contain noise $\underline{v}(k + 1)$, where $\underline{v}(k + 1)$ is a stationary Gaussian process with convariance $\underline{M}$. The measurement process is then represented by

$$\underline{z}(k + 1) = \underline{H}(k + 1)\underline{x}(k + 1) + \underline{v}(k + 1) \tag{1}$$

The Kalman filter used here is a recursive computational algorithm which remembers past data, receives future positions, and bases the estimate of the state upon a combination of past and present information.

The filter is characterised by containing a mathematical model of the ship.

In determining the value of the filter gain matrix consideration has to be given to the control vector $\underline{u}(k)$ and its associated control matrix $\underline{B}(k,k+1)$. A model of B(k,k+1) is required in the filter and the complete filter model is shown in Figure 3, leading to the overall filter equations

$$\hat{\underline{x}}(k+1) = \underline{A}(k,k+1)\hat{\underline{x}}(k) + \underline{B}(k,k+1)\underline{u}(k) + \underline{K}(k+1)\Big[\underline{z}(k+1) - \underline{H}(k+1)\{\underline{A}(k,k+1)\hat{x}(k) + \underline{B}(k,k+1)\underline{u}(k)\}\Big] \quad (2)$$

The Kalman gain matrix $\underline{K}(k+1)$ is obtained first by calculating the predicted error covariance matrix given by

$$\underline{P}(k+1,k) = \underline{A}(k+1,k)\underline{P}(k/k)\underline{A}^T(k+1,k) + \underline{C}(k+1,k)\underline{N}(k)\underline{C}^T(k+1,k) \quad (3)$$

for some initial error covariance $\underline{P}(k/k)$, and then calculating the Kalman filter gain from

$$\underline{K}(k+1) = \underline{P}(k+1/k)H^T(k+1)\Big[\underline{H}(k+1)\underline{P}(k+1/k)H^T(k+1) + \underline{M}(k+1)\Big]^{-1} \quad (4)$$

Finally the error covariance matrix is obtained using

$$\underline{P}(k+1/k+1) = \Big[\underline{I}-\underline{K}(k+1)\underline{H}(k+1)\Big]\underline{P}(k+1/k) \quad (5)$$

The above equations are used iteratively to obtain the state estimate at any future sampling time, given the initial state and error covariance, together with the measurement and disturbance noise covariances $\underline{M}$ and $\underline{N}$, the state, disturbance and control matrices, and the measurement matrix. Figure 3 gives an overall block diagram of the optimal filter.

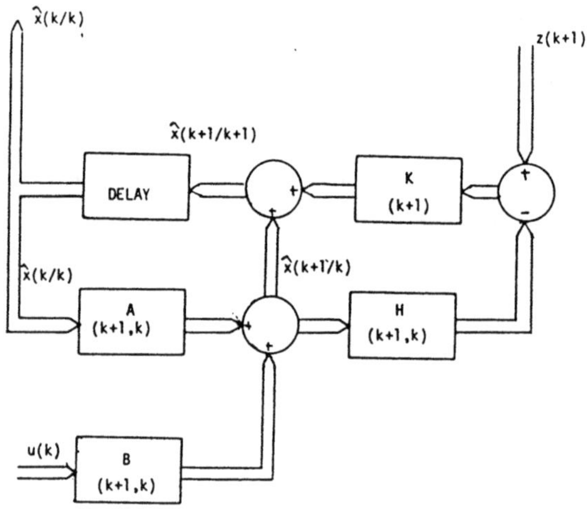

Figure 3. The optimal filter

As the ship is a non-linear system, the mathematical model used in the filter must be non-linear. It is assumed throughout that the course and speed of the vessel are constant during each sample time, with the new values being calculated during each sample period. This assumption allows the linear Kalman filter theory to apply, but it does mean that the transition matrices and filter gains have to be recalculated during each sample interval.

The covariance matrix for the measurement noise is obtained from the standard deviations of the sensors used in the various tests. For the computer simulations it is assumed that a rudder angle indicator and revolution counter are available, together with a hyperbolic position fixing system, a Doppler log to measure forward and lateral speeds, with a gyro compass and rate gyro to give heading and angular velocity. The measurement noise is assumed to have zero mean. Random number generator subroutines are used to obtain the measurement and disturbance random noise values used in the simulation.

There are two critical factors in the design of the optimal filter, firstly the modelling of the filter itself, ie, how good is the model of the ship used in the filter, and secondly the values calculated for the matrix $\underline{K}(k+1)$. The mathematical model used in the filter software is derived from the ship's hydrodynamic coefficients, which are obtained from published ship data, or, in the case of the physical model, by undertaking tank tests. Subroutines are then used to calculate the transition matrices from the data.

Computer Simulation

The vessel chosen for the computer simulations is based upon a physical scale model of a twin screw car ferry. The computer model is thus based upon the non-dimensionalised hydrodynamic coefficients derived by carrying out tank tests on a physical model. These are then scaled up appropriately to represent a full sized car ferry using the Port of Plymouth.

The Optimal Filter takes as inputs the measured values of the state vector, $\underline{z}(k+1)$, together with the previous values of the control vector, $\underline{u}(k)$. It produces a best estimate of the state vector $\hat{\underline{x}}(k+1/k+1)$ which then becomes the input to the controller. This in turn provides the demanded values of rudder angle and engine revolutions to guide the vessel automatically along a pre-defined track stored in the computer memory. The position of the vessel, together with her heading and speed are thus controlled simultaneously and automatically. Essentially there are three modes of operation to be considered. Knowledge of the statistical nature of the measurement errors together with data relating to wind and tide are used in the Kalman Filter to provide best estimates of the state vector. This is the navigation mode where the system is being used to provide the operator with more accurate position and velocity data than he would expect from using the individual measurement systems on their own. This information can be displayed upon a graphics terminal on the bridge, or at any remote position or it may be fed directly to the digital controller, which compares the estimated values with data stored in the computer memory and computes the necessary control in terms of rudder action and/or engine activity to minimise the errors. This is the fully automatic track keeping mode.

A further mode of operation would involve an automatic hazard avoidance system so that the computer automatically assesses the risk of collision with other vessels and passes the appropriate instructions to the controller so that the correct avoiding action can be taken. This final mode of operation is the subject of other research projects in Plymouth Polytechnic, Davis, Dove and Stockel, (7), Colley, Curtis and Stockel, (8).

The recommended track for deep draft vessels entering Plymouth is selected as a design specification for the project. Since the vessel is to be automatically piloted along the predetermined path, this implies that a track controller is to be used. In fact the optimal controller is both a track and a heading controller. As these two requirements can at times conflict, the optimal weightings are such that the track control dominates, except at times when an alteration of course becomes necessary, when the weightings are changed so that the heading control predominates.

Figure 4 shows a typical passage in to the port under daylight conditions. The simulation assumes that position data is being taken from the Decca Navigator using a standard deviation of 25 metres. Figure 5 shows an increase in standard deviation to 200 metres. In each case the true and filtered tracks are almost coincident, whilst the unfiltered measured data is indicated by the random line.

Figure 6 illustrates the use of the filter when one or more of the

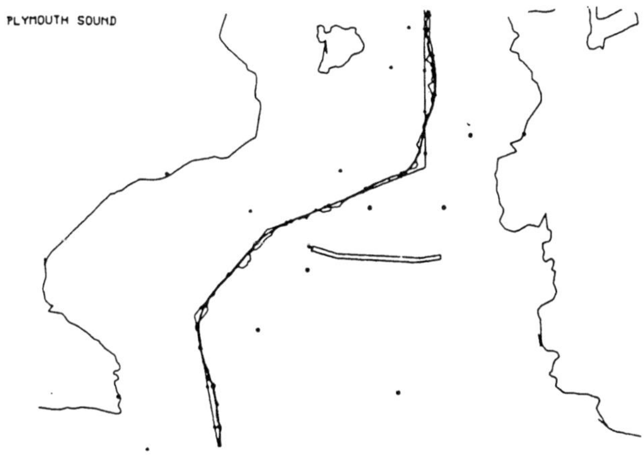

Figure 4. Daylight approach to Plymouth

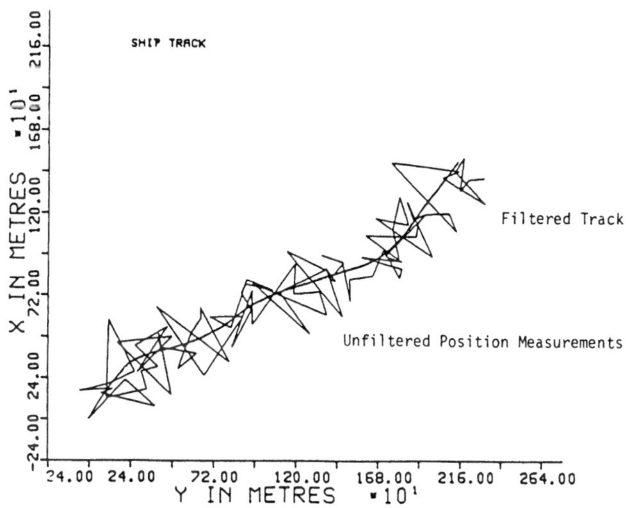

Figure 5. Filtered track at night

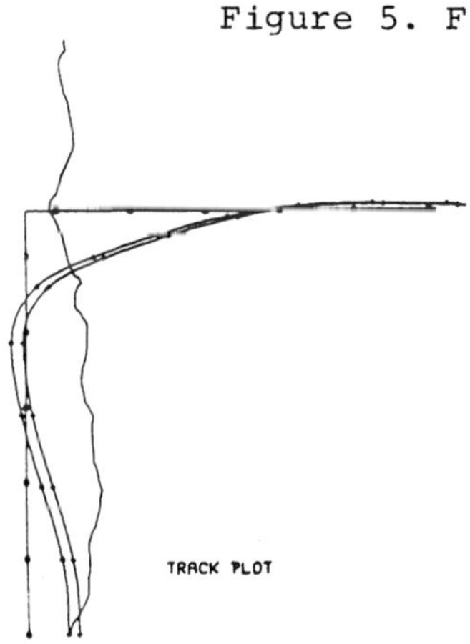

Figure 6.
Gyro compass malfunction

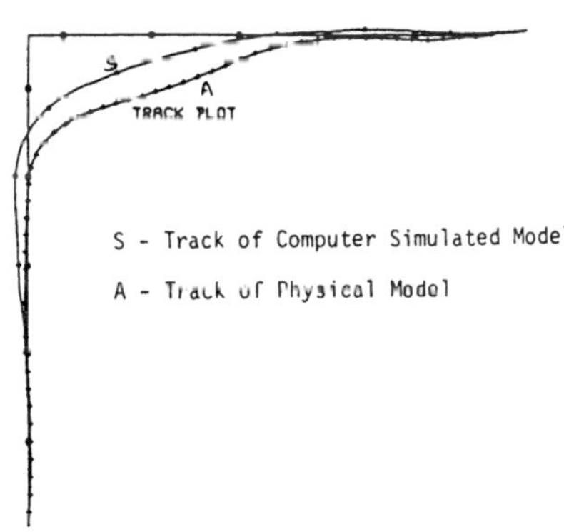

Figure 7.
Comparison of simulated and actual model tracks

measurement sub-systems ceases to function. The filter continues to estimate the state, and the vessel continues along her correct track, although the gyro compass is malfunctioning. In Figure 7 a comparison between the computer simulation of the physical model and its actual test track is shown.

Conclusions

A comprehensive digital simulation of a ship's dynamics has been set up and used to observe the time domain response of a ship in the approaches to a port when the associated track control system employed an optimal digital estimator/filter in conjunction with an optimal controller. The simulation has been used in the design of an optimal filter for installation in a physical model of a cary ferry. Tests undertaken with the physical model then confirmed the results obtained in the digital simulations, leading to a proposed automatic guidance system for use in the approaches to a port. Use of this system makes it possible to improve the safety standards in the approaches to a port particularly in conditions of bad weather, so that it is possible to enter harbour in conditions when the prudent Master would hitherto have remained "hove to" outside the port limits. In the case of a car ferry, this would improve the service offered to the passenger and enable already tight schedules to be adhered to more efficiently.

Throughout the research it has been assumed that the ship was under automatic control using a closed loop feedback system. Operated purely in the open loop navigation mode, using a digital display to give along-track and off-track positions and velocities together with an analogue display to show ownship's position relative to the surroundings (and other ships), data would be continuously available to the Master. This provides an important addition to the safety of the ship operating in restricted waters and narrow waterways.

Notation

a) Matrices and Vectors

$\underline{A}$	Discrete State Transition Matrix
$\underline{B}$	Discrete Control Matrix
$\underline{C}$	Discrete Disturbance Matrix
$\underline{H}$	Measurement Matrix
$\underline{K}$	Kalman Gain Matrix
$\underline{M}$	Covariance of Measurement Noise
$\underline{N}$	Covariance of Disturbance Noise
$\underline{P}$	State Error Covariance Matrix
$\underline{Q}$	State Error Weighting Matrix
$\underline{r}$	Desired State Vector
$\underline{S}$	Feedback Gain Matrix
$\underline{u}$	Control Vector
$\underline{v}$	Noise Vector
$\underline{w}$	Disturbance Vector
$\underline{x}$	State Vector
$\hat{\underline{x}}$	Best Estimate of State Vector
$\underline{z}$	Measured State Vector

b) Scalar Symbols

n_A, n_D	Actual and Demanded Engine Speeds (rad/s)
$r, \dot{r}$	Angular Velocity and Acceleration of Ship about Z Axis
T	Sampling Time Interval (s)
t	Time (s)
T_N	Time constant of Main Engines (s)
T_R	Time constant of Rudder Servo (s)
U	Track Velocity (m/s)
u	Forward Velocity of Ship (m/s)
v	Lateral Velocity of Ship (m/s)
x,y,z	Ship Related Cartesian Co-ordinates (m)
x_0, y_0, z_0	Earth Related Cartesian Co-ordinates (m)
δ_A, δ_D	Actual and Demanded Rudder Angles (rad)
ρ	Density of Water (kg/m^3)
ψ_A, ψ_D	Actual and Demanded Heading of Ship (rad)

References

1. Dove, M J:
   Automatic Control of Large Ships in Pilotage and Berthing.
   J. Nav. Vol 27, No 4 (1974), pp 519 - 529

2. Burns, R S, Dove, M J, Bouncer, T H and Stockel, C T:
   A Discrete, Time Varying, Non-Linear Mathematical Model for the Simulation of Ship Manoeuvres.
   First Intercontinental Maritime Simulation Symposium and Mathematical Modelling Workshop, Munich, W Germany. June 1985

3. Sage, G F and Luse J D:
   Integration of Transit, Omega and Loran-C for Marine Navigation.
   Navigation (USA), Vol 30, No 1 (1983)

4. Grimble, M J, Patton, R J and Wise, D A:
   Use of Kalman Filtering Techniques in Dynamic Ship Positioning Systems.
   Proc. IEE Vol 127, Pt.D No 3 May 1980, pp 93-102

5. Healey, M and MacKinnon, D J:
   Design of a Digital Self-Alignment Controller for a Ship's Inertial Navigation System at Sea.
   Proc. IEE Vol 122, No 1 January 1975

6. Kalman, R E and Bucy, R S:
   New Results in Linear Filtering and Prediction Theory
   J. Basic Eng. March 1961

7. Davis, P V, Dove, M J and Stockel, C T:
   A Computer Simulation of Multi-Ship Encounters
   J. Nav. Vol 35, (1982) pp 347-352

8. Colley, B A, Curtis, R G and Stockel, C T:
   On Marine Traffic Flow and Collision Avoidance Computer Simulation
   J. Nav. Vol 37, No 2 (1984) pp 232-250

Session 7 Marine Maneuvering and Motion Simulation

Marine Maneuvering Simulation

Max H. CARPENTER
MITAGS
5700 Hammonds Ferry Rd.
Linthicum Hghts, MD 21090

ABSTRACT

To provide guidance to any potential buyer of a major Marine Manuevering Simulator, the following outline of recommendations and specifications is offered. The items included could, considering any vessel characteristics, satisfy in terms of "feel and visual accuracy", the fidelity that ship's simulation demands.

INTRODUCTION

The constant search for an optimum method of imparting shiphandling knowledge and skill to the neophyte mariner has finally lead to the use of electronic simulation for training. It is remarkable that while the sea is one of the oldest means of heavy goods and passenger carriage, the industry it spawned was one of the last to embrace simulation as an acceptable means of training. In some respects the simulator industry was to blame for this slow acceptance. Steamship operators were reluntant to move toward use of these devices due, in part, to the lack of substantative information on which to base their decisions. Even now, there is a wide divergence in the way this training technology is applied. There are many areas of the world where the shipping industry contributes significantly to the economy of that country, yet there is no change in their traditional ways of training the officers and men involved. At the moment, the number of facilities throughout the world with advanced maritime training simulators is astonishing small. (The International Marine Simulator Forum (IMSF) has published a list showing the location of these devices.)

As the size of the modern vessel increases, the number of personnel involved with its handling at sea tends to decrease. With this trend, it is almost certain that the training for the handling of these types of vessels must be done by use of a simulator. It is obvious that the costs of operation plus the risk involved in manning with officers unfamiliar with the vessel makes it impracticable to train at sea.

Most regulatory bodies throughout the world prescribe a length of sea time that must be served by a junior officer before the next grade can be achieved. The amount of sailing time available to satisfy these regulations is being drastically reduced by the many factors affecting modern shipping. However, the U.S. Coast Guard, as one of these regulators, has recognized this problem and has changed their rules and now permit some of this required time to be accumulated by training in an approved ship maneuvering simulator.

The fidelity of any simulator results initially from the accuracy and response of the instrumentation coupled with the simulated visual environment to provide the cues that when integrated by the student officer make up the vessel's "feel".

Because the visual input to the trainee is so important to the success of a full mission ship simulator, it will be dealt with first.

THE VISUAL SYSTEM

Two distinct classifications of visual systems are recognized.

(1) Pure nocturnal: This is a display in which the port facilities, landmass and any other ships involved in the exercise are displayed as a pattern of lights. In this system, the computor controlled spot projects form the scene with a resolution of at least two minutes of arc. The pattern of lights accurately represent, in both color and intensity, objects as seen from a darkened bridge on a completely dark night. In these systems, usually no profile or silhouette of the objects are visible.

(2) Daylight: There are several different versions of the daylight system. The most popular is the Computor Generated Image (CGI). A video based system which shows a full pictorial display, usually in color and can include port facilities,

landmass and other ships. Another scheme uses a detailed model board over which a video camera travels on a gantry under computor control. The resulting images of these two systems are projected by a multiple of video projectors depending on the desired horizontal field of view. This type of visual has been used to generate a nocturnal scene and has not been very successful, which is in part due to the limited brightness range of the raster and the restricted resolution of the video generated light spots. Another scheme that has been tried is a continuous film device. While this provides excellent resolution, preparing the data base and finally projecting the finished product is a difficult procedure. Systems such as this are becoming outmoded as progress continues in projection techniques that will provide bright high resolution video. In all of the above methods, the bridge illumination must be subdued otherwise it will wash out the simulated daylight scene.

To assure the highest quality of simulation, no matter which visual system is used, the following items must be considered:

(1) Depth of field.

In essence, depth of field can be considered as perceived range to the object or the apparent distance. Several different techniques can be used to enhance the apparent depth of field of a visual system. A simple method of achieving this effect is to locate the resolving surface of the visual system at a distance which defeats the natural ranging ability of the human eye. In most cases, this distance is in excess of 27 feet. Depth of field can be simulated by a device called a beam splitter which makes use of the virtual image phenomena to increase the apparent distance to the object from the viewer's eye. The construction of the elements of this system limit its horizontal field of view. Attempts to extend it into angles in excess of 180° have not yet been successful. As the interest in good maritime visuals increase, the problems of construction will, no doubt, be overcome. Use of long flexible plastic sheets for the primary and secondary surfaces of the lens and formed to give a wide horizontal angle might hold promise for the beam splitter approach. However, many of the simulators in current use do not even attempt to enhance the illusion of range; though realism suffers as a result.

(2) Field of View. (FOV)

An uninterrupted 360 degree field of view is currently available only in a nocturnal presentation for maritime simulation. At the present time there is no continuous and undisturbed wide angle F.O.V. in use for daylight maritime simulation. Many simulators that use a Computor Generated Image and video projectors to produce the scene, must still piece together rasters to achieve the desired field of view. The disadvantage to this approach is that artifacts are produced in the picture by the joining of the raster edges. The number of projectors used depend on the desired horizontal angle. Physically, it is impossible to locate all of these projectors at the optical center of the simulator theatre; therefore, parallax is another problem facing the present marine simulator designer seeking wide F.O.V. solutions.

With these optical design questions looking for an economically feasible answer, perhaps some of the new laser video projection systems used in conjunction with a spinning optical head might prove promising in eliminating both parallax and unwanted artifacts in the scene.

(3) Vertical Angle.

The vertical angle of the scene should be approximately 30 degrees. That is; plus or minus 15 degrees from the horizon. This foreground visual angle is needed if the simulator is to provide situations such as docking and other maneuvers where own ship and another vessel might be close together. To complete the illusion, own ships forebody should be viewable from the simulator. It should be possible to select the appropriate image of the vessel being simulated. In many current simulators, this view is projected on the same screen as the other environmental details. Using this technique, parallax will cause own ships' forebody to "windshield wipe" back and forth as the viewer changes position on the bridge. With enough space between the main screen and the bridge, an inclined auxillary screen can be used to project the ownship image thus reducing or eliminating the undesirable effect.

(4) Resolution.

The need for resolution is always challenged because of its cost. If it could be available at an affordable price, no-one would object to having it included in their simulator. The problem might be how to specify it amidst the plethora of "pixels and edges" that are tossed about by the computor generated imagery designer or programmer while describing picture detail. Quite simply, resolution can be described as minutes of arc. A totally plausable number to specify is two minutes of arc which means approximately eleven thousand separate and distinct light spots edge to edge on a 360 degree horizon circle.

(5) Luminance and screens.

Luminance describes the brightness on a resolving surface or screen and is spoken of as light intensity emitted per square centimeter of illuminated surface. Many of the existing simulators use a standard fabric screen for either front or rear projection. When stretched on a frame into a cylinder, this technique is entirely satisfactory for a wide field of view. If the screen requirement involves a sphere or a truncated sphere for the reflecting surface, quite logically one might look for "gain" from the coating they plan to use. This is particularly true if 220 degrees or more are to be illuminated by the projectors. As gain usually also means narrow bandwidth, consequently the unwanted reflection or cross coupling from one side of the screen to the other could, by proper design, be reduced. If you should require "gain" from your screen, be advised it now is necessary to provide a perfect surface that has almost no pertubations. Otherwise, when your screen is illuminated every defect becomes an unwanted artifact in the scene. If this is the problem, then your solution is to forget gain and use a matte finish as your final coat.

THE BRIDGE LAYOUT

The layout of the bridge is quite often treated as the most important part of the simulator. Very likely it is because this is the area of the simulator where the customer/user is familiar with the equipment that is to be included. However, the intended use of the simulator must be considered to assure that proper instrumentation and controls are available without overloading the bridge with unnecessary items. Of course, care needs to be exercised that the layout abets the visual scene by providing a logical and effective operating area for the trainee to occupy. This means traditional conning positions should be considered along with the possibility of new configurations in the future. The shape and size of the bridge usually is a function of the visual system specifications. Unless the bridge is to have motion; that is roll, pitch and heave, the consoles and the steering stand can be left movable so bridge arrangement can be altered when appropriate. To reduce unwanted reflections, many of the existing simulators do not use glass in the bridge windows. Sometimes this is not possible due to noise or the lack of cooling or heating outside the wheelhouse. There are three important points on which to focus when settling on your simulator bridge design.

(1) The impact the layout will have on viewing the visual scene.

(2) Does the layout relate to the types of vessels to be simulated in the foreseeable future?

(3) Will the layout support either bridge team or individual training?

INSTRUCTOR CONTROL AREA.

If the visual system is the most important part of a marine simulator, the very next item in the hierarchy is the Instructor Control Area. The one essential that must be designed into a training simulator is ease of operation by one instructor. This, of course, has to be done without degrading the quality of training. To be effective, the instructor must have available communications, graphics, and data displays that will permit him to interface with the trainees in many modes. For example, training exercises usually require interaction in which the instructor plays a number of different parts. For example; he may be the captain of another vessel or tug or he may be the chief engineer or any one of the positions aboard ship that are manned while maneuvering. As technologically advanced as the equipment may be, one of the most effective aspects of any simulator is the role playing by the instructor. While involved in this activity, he must also be able to accumulate and evaluate data regarding the trainees performance. This data should also be available in a "hard copy" form that can be used when debriefing the trainees following an exercise. Controls should be provided in this area so that when desired, to change the exercise parameters readily simple keyboard commands or push-button controls should be provided. When desired, the instructor should be able to alter the characteristics of the simulated through data manipulation. A simple example of what might be required is to change an "in ballast" vessel to a "loaded" configuration during a particular exercise. The instructor will, from time to time, wish to alter underwater topography, and add other contact vessels. In other words, the instructor should easily be able to accomplish a wide range of temporary revisions to the training program. Recently, there has been much talk about using interactive modes in training. This is where two or more simulators are "hooked up" so that the actions of the trainee on one of the bridges or training positions affects the response of the trainees involved. While this seems interesting, as a training exercise, it overloads the instructor. By inserting these additional uncontrolled variables into the evaluation process, it becomes difficult to grade the individual trainees. To monitor the activity on the bridge during an exercise, a simple intercom that is normally in a "listen" mode, will suffice. Visual observation either by television or direct view does not seem to be necessary. Remember, keep your instructor involved in the job he is best suited to do.

DOCUMENTATION:

While, in most cases, it receives little or no attention, an essential part of any successful training device is the support documentation. Perhaps in relation to the overall procurement, documentation is not important to the equipment manufacture, but during the life of the simulator, this item is of great importance to the end user. The moment of truth comes when the instructor tries to make simple modifications to the exercises that were supplied by the manufacturer.

Basically, the documentation should provide the information necessary for maintaining, modifying and upgrading and to operate the system. In the past, some simulators have required a specialist just to operate it and make simple changes to its programs. There is no justification for making control of the device that demanding. Most instructors are highly competent shiphandlers and should not be wasting their valuable time deciphering complex instructions. The documentation should support all of the functions available to him through the instructors' console. In addition, it should allow him to make needed changes to the disc files. For example; additons or corrections to harbor facilities as they occur or simple revisions to the form used in recording evaluation data. The form that documentation should take is highlighted by these key words, simplicity and ease of use. In the final analysis, the "full mission" simulator with its more powerful computor, will be more flexible than its smaller cousin, the micro computor driven desk top simulator. There is no reason why its documentation, while easy to use, should not also be as powerful.

References

1. Diane M. Gayeski & David V. Williams-- Interactive Media-- (Prentiss Hall, 1985)

2. D.Dey, U. Hohage & Manfred Schmidt--Models for Shiphandling Simulators--MBB, Bremen, Germany, 1980.

3. T.J. Hammell--The Training Device is More Than a Simulator-Marsim Proceedings, 1981.

4. Max H. Carpenter--How Much Simulator is Really Needed in Maritime Training--SCS Simulation Technical Journal,1984.

5. Joseph J. Puglisi--An Overview of IMSF--SCS Conference, Norfolk, Va.

A Discrete, Time Varying, Non-Linear Mathematical Model for the Simulation of Ship Manoeuvres

R. S. Burns, M. J. Dove, T. H. Bouncer, C. T. Stockel

Plymouth Polytechnic.
U.K.

Summary

This paper considers the development of a discrete, time-varying non-linear mathematical model for a ship based upon eight state variables. There are two deterministic inputs - demanded rudder and engine speed plus four stochastic inputs in the form of wind and current vectors. The model is validated by conducting a comprehensive simulation study using known linear and non-linear hydrodynamic coefficients for a full-size vessel. A comparative evaluation of simulated results against real data is undertaken for a set of standard manoeuvres. It is demonstrated that a non-linear, time-varying multivariable model accurately describes the motion of the hull, particularly in tight manoeuvres.

Introduction

A ship may be considered to be a multivariable system with many inputs and outputs. If a true representation of its manoeuvring characteristics is to be achieved, care must be taken over the selection of system parameters and the manner in which they change during a particular manoeuvre.

This paper considers the development of a discrete, time-varying non-linear mathematical model of a ship, Burns (1), Dove (2), Burns et al (3). Multivariable system theory is employed to construct a model based upon eight state variables, namely forward and lateral position and velocity, heading, yaw-rate, rudder angle and engine speed. The model has two deterministic inputs - demanded rudder and engine plus four stochastic disturbance inputs in the form of wind and current vectors. The response of the model to these latter inputs is not considered in this paper.

Equations of Motion

The ship is considered to be a rigid body with three degrees of freedom, in surge, sway and yaw. Ship motions in the other three degrees of freedom, roll, pitch and heave are considered small enough to be neglected. It is convenient to describe the motion in terms of a moving system of axes coincident with the mass centre of the hull as illustrated in Figure 1. This gives rise to an Eulerian set of equations of motion which may be written in the form

$$\begin{aligned} m\dot{u} - mr\nu &= X \\ m\dot{\nu} + mur &= Y \\ I_z\dot{r} &= N \end{aligned} \qquad (1)$$

Techniques employed in obtaining expressions for hydrodynamic forces and moments are well covered in the literature Abkowitz (4) and the usual method is to apply a Taylor series expansion. For applications such as course-keeping, where changes in rudder and heading angles do not usually exceed five degrees, a linear approximation, using only the first order terms in the expansion, is normally quite adequate. In a track-keeping situation where large changes in heading can be expected, it becomes necessary to include second and third order expansion terms.

Surge equation. The complete surge equation in dimensionalised form is

$$\begin{aligned} m\dot{u} - mr\nu = {} & X_{\dot{u}}\dot{u} + X_u(u + u_c) + \overline{X}_{uu}u^2 + \overline{X}_{uuu}u^3 + \overline{X}_{\nu\nu}\nu^2 \\ & + \overline{X}_{rr}r^2 + \overline{X}_{\delta\delta}\delta_A^2 + \overline{X}_{uu}un_A + \overline{X}_{nn}n_A^2 \\ & + X_{u_a}u_a \end{aligned} \qquad (2)$$

In the above equation a shorthand subscript and bar notation has been adopted, for instance

$$X_u = \frac{\partial X}{\partial u}, \quad \overline{X}_{uu} = \tfrac{1}{2} X_{uu} = \tfrac{1}{2}\frac{\partial^2 X}{\partial u^2}$$

The dimensionalised hydrodynamic coefficients are obtained from the non-dimensional values in the usual manner

$$X_u = (\tfrac{1}{2}\rho L^2 U)X_u'$$

Sway and Yaw Equations. The dimensionalised sway and yaw equations are

$$m\dot{\nu} + mur = Y_{\dot{\nu}}\dot{\nu} + Y_\nu(\nu + \nu_c) + Y_{\dot{r}}\dot{r} + Y_r r + \bar{Y}_{nn}n_A^2 + \bar{Y}_{\nu\nu\nu}\nu^3$$

$$+ \bar{Y}_{r\nu\nu}r\nu^2 + \bar{Y}_{nn\delta}n_A^2\delta_A + \bar{Y}_{nn\delta\delta\delta}n_A^2\delta_A^3$$

$$+ \bar{Y}_{\delta\nu\nu}\delta_A\nu^2 + Y_{\nu_a}\nu_a \qquad (3)$$

$$I_z\dot{r} = N_{\dot{\nu}}\dot{\nu} + N_\nu(\nu + \nu_c) + N_{\dot{r}}\dot{r} + \bar{Y}_{nn}n_A^2 + \bar{N}_{\nu\nu\nu}\nu^3 + N_r r$$

$$+ \bar{N}_{r\nu\nu}r\nu^2 + \bar{N}_{nn\delta}n_A^2\delta_A + \bar{N}_{nn\delta\delta\delta}n_A^2\delta_A^3$$

$$+ \bar{N}_{\delta\nu\nu}\delta_A\nu^2 + Y_{\nu_a}\nu_a \qquad (4)$$

Steering Gear and Main Engine. These are both modelled by first order linear differential equations.

$$\dot{\delta A} = \frac{1}{T_R}\delta_D - \frac{1}{T_R}\delta A \qquad (5)$$

$$\dot{n}_A = \frac{1}{T_N}n_D - \frac{1}{T_N}n_A \qquad (6)$$

Where δ_D and n_D are the demanded rudder angle and demanded engine speed respectively.

State Space Formulation

Much attention was devoted to the choice of state variables and the state vector was finally based on the ship body axes.

$$X^{T} = (\delta_A \quad n_A \quad x \quad u \quad y \quad v \quad \psi \quad r) \tag{7}$$

This state is affected by the forcing vector

$$U^{T} = (\delta_D \quad n_D \quad u_c \quad \nu_c \quad u_a \quad \nu_a) \tag{8}$$

Equations (5), (6), (2), (3) and (4) can be arranged in the following matrix form.

$$\frac{d}{dt}\begin{bmatrix}\delta_A\\ n_A\\ x\\ u\\ y\\ \nu\\ \psi\\ r\end{bmatrix} = \begin{bmatrix}-\frac{1}{T_R} & 0 & 0 & 0 & 0 & 0 & 0 & 0\\ 0 & -\frac{1}{T_N} & 0 & 0 & 0 & 0 & 0 & 0\\ 0 & 0 & 0 & 1 & 0 & 0 & 0 & 0\\ f_{41} & f_{42} & 0 & f_{44} & 0 & f_{46} & 0 & f_{48}\\ 0 & 0 & 0 & 0 & 0 & 1 & 0 & 0\\ f_{61} & f_{62} & 0 & f_{64} & 0 & f_{66} & 0 & f_{68}\\ 0 & 0 & 0 & 0 & 0 & 0 & 0 & 1\\ f_{81} & f_{82} & 0 & f_{84} & 0 & f_{86} & 0 & f_{88}\end{bmatrix}\begin{bmatrix}\delta_A\\ n_A\\ x\\ u\\ y\\ \nu\\ \psi\\ r\end{bmatrix}$$

$$+\begin{bmatrix}\frac{1}{T_R} & 0 & 0 & 0 & 0 & 0\\ 0 & \frac{1}{T_N} & 0 & 0 & 0 & 0\\ 0 & 0 & 0 & 0 & 0 & 0\\ 0 & 0 & g_{43} & 0 & g_{45} & 0\\ 0 & 0 & 0 & 0 & 0 & 0\\ 0 & 0 & 0 & g_{64} & 0 & g_{66}\\ 0 & 0 & 0 & 0 & 0 & 0\\ 0 & 0 & 0 & g_{84} & 0 & g_{86}\end{bmatrix}\begin{bmatrix}\delta_D\\ n_D\\ u_c\\ \nu_c\\ u_a\\ \nu_a\end{bmatrix} \tag{9}$$

The coefficients f_{41} , f_{42} ... g_{84}, g_{86} are all time-varying and so, for example

$$f_{41} = \frac{\overline{X}_{\delta\delta}\cdot\delta_A}{m - X_{\dot{u}}} = \frac{(\frac{1}{2}\rho L^2 U).\overline{X}'_{\delta\delta}\,\delta_A}{m - X_{\dot{u}}}$$

f_{41}, therefore, is a function of the instantaneous total velocity U and rudder angle δ_A.

Equation set (9) represent the time-varying state equations for the ship and are expressed by the state matrix vector differential equation

$$\dot{X}(t) = F(t)X(t) + G(t)U(t) \tag{10}$$

It is convenient to partition the G matrix in terms of the control forcing function δ_A and n_A and the disturbance forcing functions u_c, ν_c, u_a and ν_a so that

$$\dot{X}(t) = F(t)X(t) + G_C(t)U(t) + G_D(t)W(t) \tag{11}$$

The corresponding discrete solution is

$$X((K + 1)T) = A(T, KT)X(KT) + B(T, KT)U(KT)$$

$$+ C(T,KT)W(KT) \tag{12}$$

where

$$A(T,KT) = e^{F(t)T}$$

$$B(T,KT) = (e^{F(t)T} - I)\ F(t)^{-1}G_C(t)$$

$$C(T,KT) = (e^{F(t)T} - I)\ F(t)^{-1}G_D(t) \tag{13}$$

Mathematical Model Validation

The ability of the mathematical model to accurately describe the motion of a vessel during tight manoeuvres was assessed by conducting a series of tests with the equations of motion (2) to (4) expressed in the following formats:

(a) Linear Model

The F and G matrices are time-invariant and comprise only of the linear coefficients Yv, Yr, Nv, Nr, etc. Surge equations are excluded and the modelling is for a constant forward speed only.

(b) Quasi-Linear Model

The F and G matrices are a time-varying function of total velocity U, but still consist of only linear coefficients. A range of forward speeds may be simulated.

(c) Non-Linear Model

The F and G matrices are time-variant and include all linear and non-linear terms. The model is again valid over a range of forward speeds.

Study Vessel

A comparative evaluation with full-scale measurements taken by Morse and Price for the USS Compass Island (5) has been undertaken in order to select the best form of the mathematical model.

The USS Compass Island was constructed with a Mariner hull form, and a complete set of hydrodynamic coefficients for this class of vessel have been measured by Chislett and Strom-Tejsen (6) using a planar-motion mechanism.

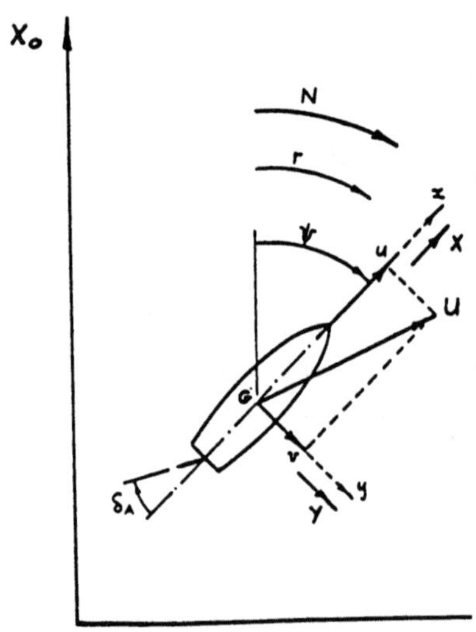

Figure 1

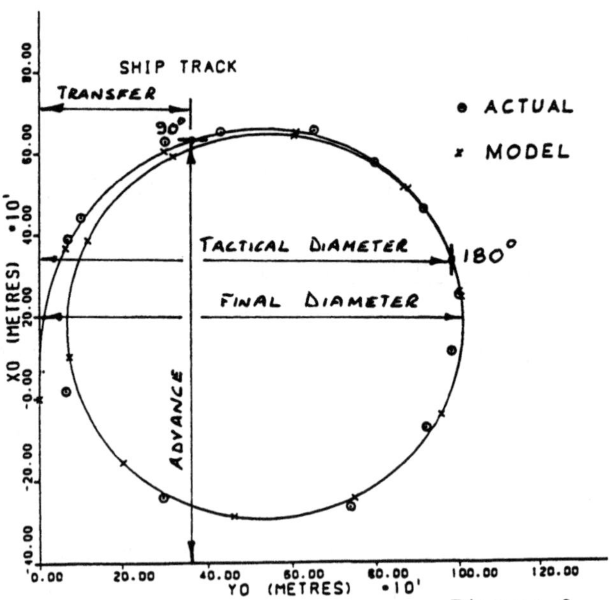

Figure 2

Manoeuvring Simulations

Turning Circle Manoeuvres

Turning circles are used to determine the effectiveness of the rudder to produce steady-state turning characteristics. Simulation runs using approach speeds of 7.717 m/s (15 knots) for all three mathematical models and 2.572 m/s (5 knots) for the quasi-linear and non-linear models only, were performed for rudder angles of ±5, 10, 15, 20, 25 and 30 degrees.

A total of 60 simulation runs were carried out and a typical turning circle is given in Figure 2. Figures 3, 4, 5 and 6 summarise the results in terms of advance, transfer, tactical diameter and final diameter as defined in Figure 2 for the two approach speeds. The results for steady-state loss in forward speed due to a turning manoeuvre are shown in Figure 7.

Dieudonne Spiral Manoeuvre

The Dieudonne spiral manoeuvre is used to provide a qualitative measure of course stability for surface ships. Steady-state yaw-rate is plotted as a function of rudder angle in a gradually increasing, and then decreasing spiral manoeuvre. Results are given in Figure 8 along with the Morse-Price data.

The simulation was performed at approach speeds of 7.717m/s and 2.572 m/s. In each case the rudder was set to +25 degrees and then decremented in steps of -5 degrees to -25 degrees (finer steps being used around the origin) and incremented back again to +25 degrees in steps of +5 degrees. At each increment the ship was allowed to settle into a steady turn and the corresponding yaw-rate noted.

Kempf Zig-Zag Manoeuvre

This manoeuvre gives an indication of the effectiveness of the rudder to initiate and check changes of heading. Results are shown in Figure 9.

The simulation was conducted at a single approach speed of

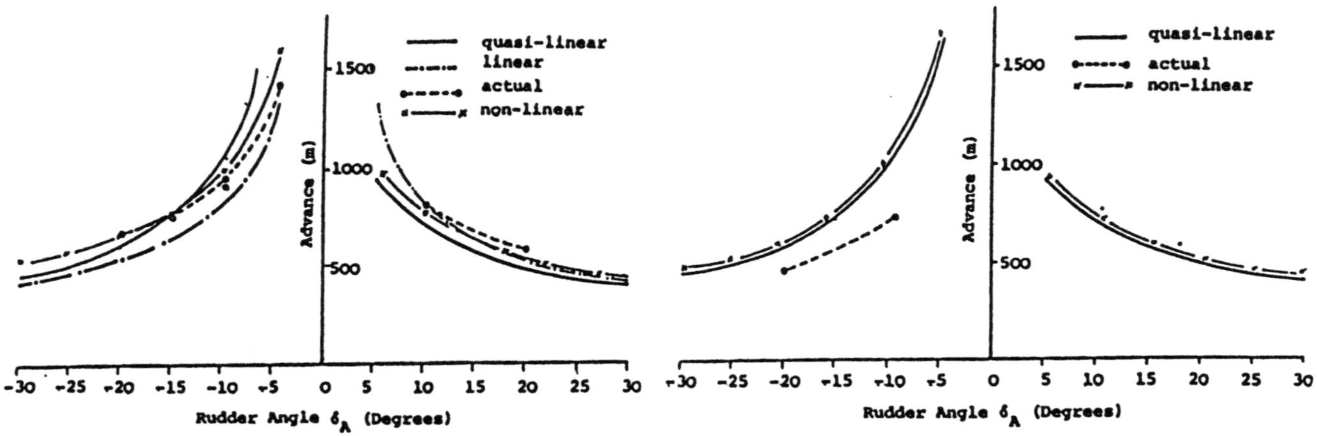

Figure 3. Advance

(a) 7.717 m/s Approach

(b) 2.572 m/s Approach

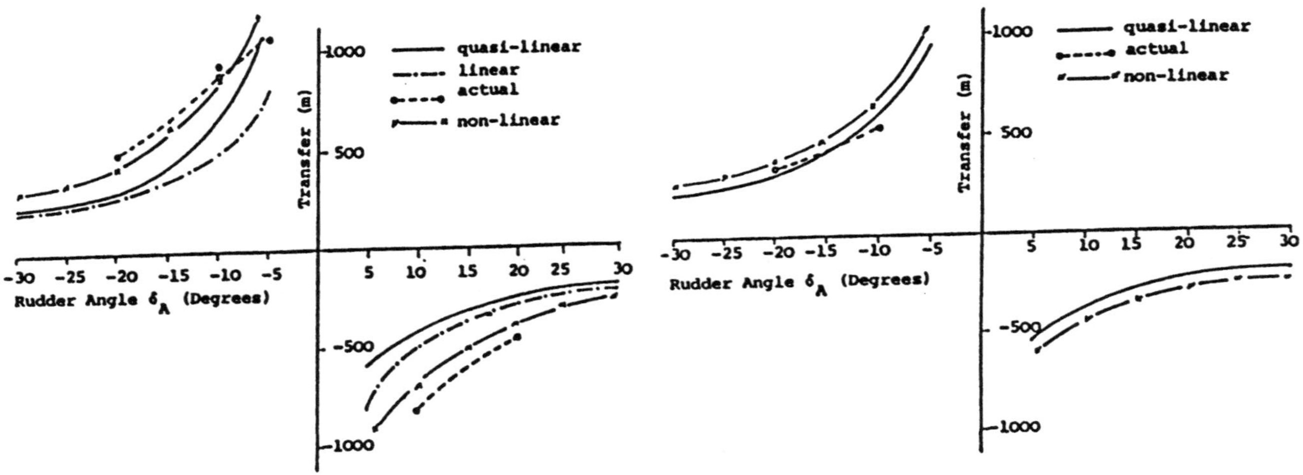

Figure 4. Transfer

(a) 7.717 m/s Approach

(b) 2.572 m/s Approach

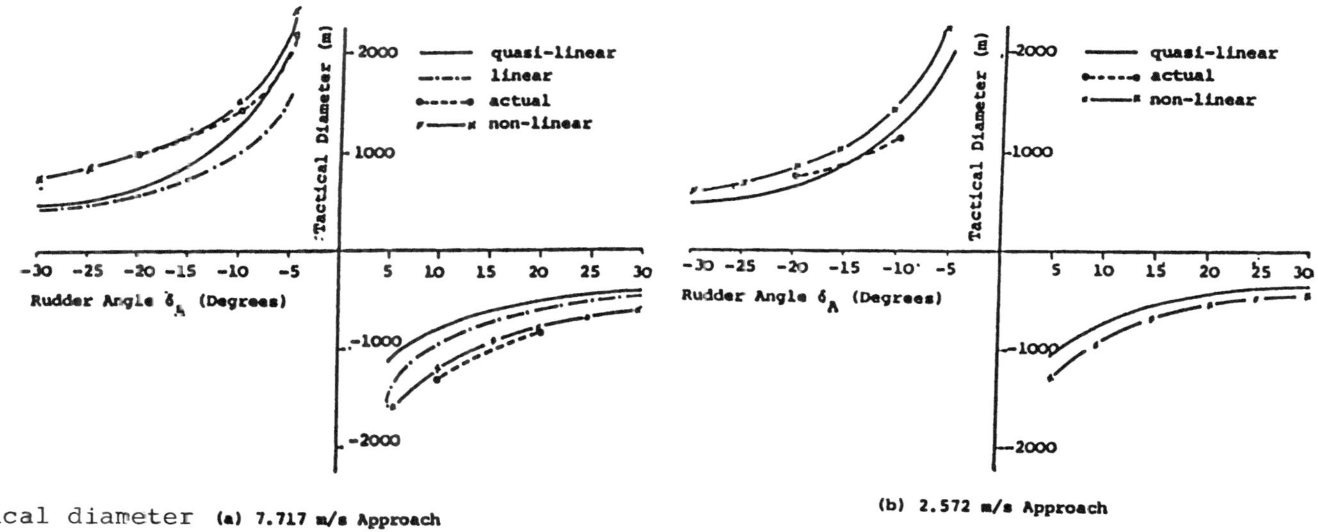

Figure 5. Tactical diameter

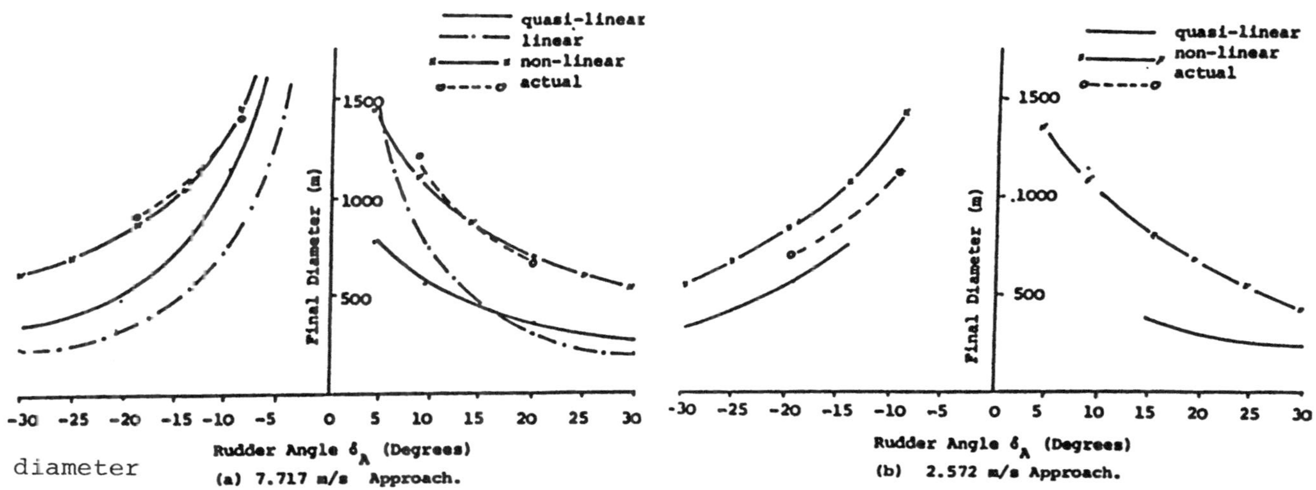

Figure 6. Final diameter

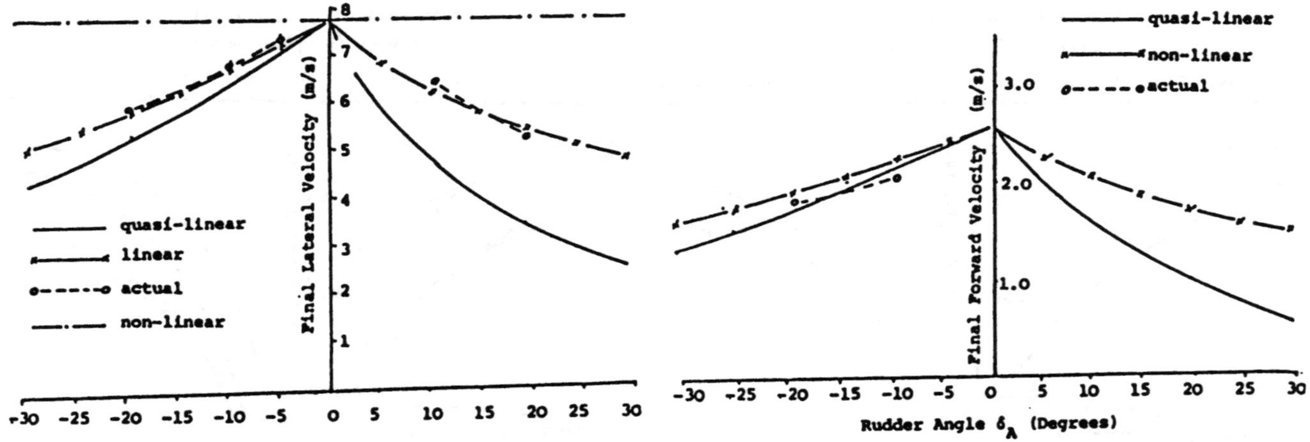

(a) Final Forward Velocity

(a) Final Forward Velocity

Figure 7.
Final forward velocity

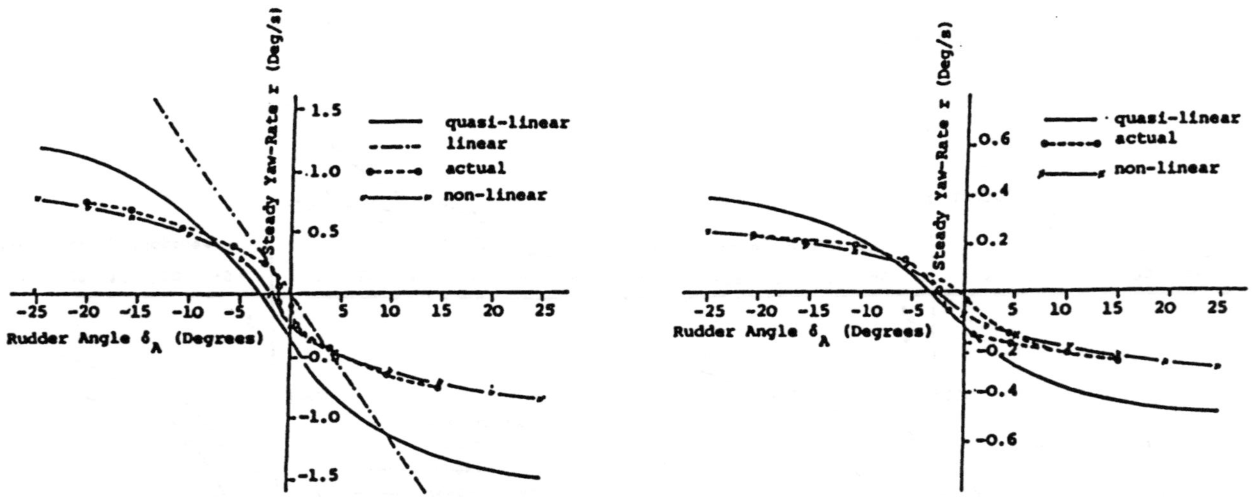

(a) 7.717 m/s Approach

(b) 2.572 m/s Approach

Figure 8. Diendonne spiral

7.717 m/s as no data for the real ship was available at the slower speed. Initially the demanded rudder was set to +20 degrees then, as the simulation continued, a check was made on the heading, and when it lay within a tolerance band of 18-22 degrees, the demanded rudder was switched to -20 degrees. The process was repeated for several changes in demanded rudder.

Starting Trials

These were performed to compare the dynamics of the vessel in the x-direction with data extracted from the Morse-Price trials. With the propeller moment and side-thrust terms removed (so that the ship would travel in a straight line) and the vessel stationary with rudder amidships, a step change in demanded engine speed was ordered. The resulting responses are shown in Figure 10.

The final steady-state relationship between forward speed and engine revolutions obtained from these trials were used as initial conditions for the turning circle, Dieudonne and Kempf manoeuvres.

Analysis of Results

The three phases of the turning circle manoeuvre, advance and transfer, tactical diameter and final diameter provide a measure of the accuracy of the mathematical models in terms of (a) initial transient, (b) final transient, (c) steady-state performance.

Figures 3 and 4 show that all the mathematical models are reasonably accurate in the initial transient, but Figures 5 and 6 reveal that towards the end of the transient and into the steady-state only the non-linear model retains its accuracy. For example, with 20 degrees of starboard ruddor, the final diameter of the real ship and the non-linear model is about 1000m, compared with about 500m for the linear and quasi-linear models. This tightness of turn in the latter case explains the excessive speed loss and lateral velocity experienced by the quasi-linear model shown in Figure 7.

The Dieudonne spiral results of Figure 8 again emphasise that only the non-linear model provides an accurate rudder to yaw-rate relationship over the rudder angle range ± 25 degrees. Note, however, that over the range ± 5 degrees, the slope (which is the gain constant K in the Nomoto model) is correct for all three mathematical models. A further by-product of this test is the conclusion that the vessel has control-fixed stability, although a close inspection of the Morse-Price data around the origin for the USS Compass Island shows a slight tendancy towards instability.

Results for the Kempf manoeuvre given in Figure 9 illustrate that, as with the turning circle, all three mathematical models produce fairly accurate results in a manoeuvre which is primarily an initial transient test.

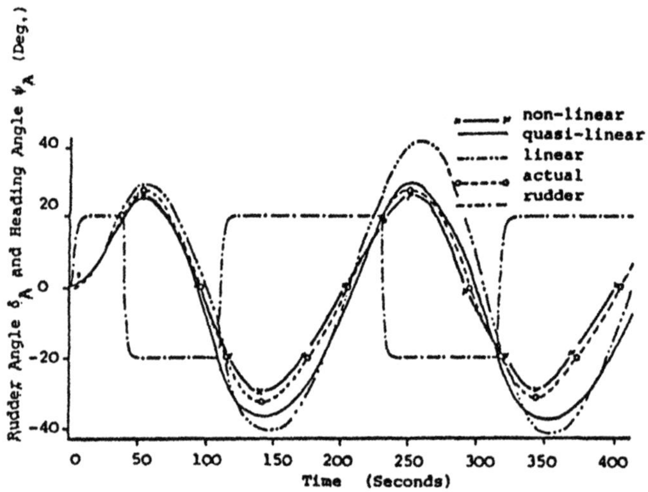

Figure 9.

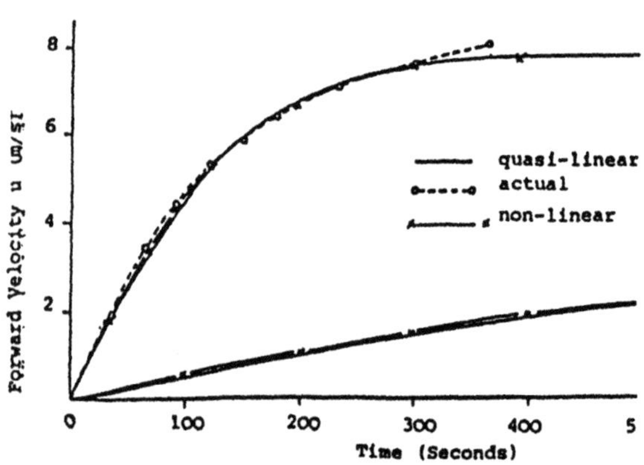

Figure 10.

Figure 10 indicates that both non-linear and quasi-linear models very closely represent the x-direction dynamics of the vessel.

Conclusions

The conclusions that must be drawn from this set of tests is that the non-linear model is the only mathematical model to accurately represent the three degrees of freedom ship motion in all manoeuvring regimes, particularly tight manoeuvres.

NOTATION

A, B	Discrete State and Control Transition Matrices
C	Discrete Disturbance Matrix
F, G	Continuous Time System and Forcing Matrices
X, U, W	State, Control and Disturbance Vectors
I_z	Moment of Inertia about z axis (kg m²)
L	Length of ship between perpendiculars (m)
m	Mass of ship (kg)
n_A, n_D	Actual and Demanded engine speeds (rad/s)
T_n, T_R	Time constant of main engines and rudder servo(s)
U	Track and forward velocities of ship (m/s)
u_a, u_c	Forward components of wind and current velocities (m/s)
ν	Lateral velocity of ship (m/s)
ν_a, ν_c	Lateral components of wind and current velocities (m/s)
x, y, z	Ship related orthogonal co-ordinates (m)
δ_A, δ_D	Actual and Demanded rudder angles (rad)
ρ	Density of water (kg/m³)
ψ	Actual heading of ship (rad)

REFERENCES

1. Burns. R. S. The Automatic Control of Large Ships in Confined Waters. Ph.D. Thesis, Plymouth Polytechnic, Sept. 1984.

2. Dove, M. J. A Digital Filter/Estimator for the Control of Large Ships in Confined Waters. Ph.D. Thesis, Plymouth Polytechnic, Oct. 1984.

3. Burns. R. S., Bouncer, T. H. and Dove, M. J. Automatic Pilotage of Large Ships in Confined Waters - A Multivariable Approach. Symposium on Application of Multivariable Systems Theory, Inst. of Meas. and Control, Oct. 1982.

4. Abkowitz, M. Lectures on Ship Hydrodynamics, Steering and Manoeuvrability. Hy-A Report, Hy, 5., Denmark, 1964.

5. Morse, R. V. and Price, D. Manoeuvring Characteristics of the MARINER Type Ship (USS Compass Island) in Calm Seas. Sperry Polaris Management, Sperry Gyroscope Company, New York, Dec. 1961.

6. Chislett, M. S. and Strom-Tejsen. Planar Motion Mechanism Tests and Full-Scale Steering and Manoeuvring Predictions for a MARINER Class Vessel. International Shipbuilding Progress, Vol. 12, p. 201-224, May 1965.

Modeling and Simulation of the Roll Motions of a Ship

J. van Amerongen
Control Group, Electrical Engineering Department
Delft University of Technology
P.O.box 5031, 2600 GA Delft, The Netherlands

and P.G.M. van der Klugt
van Rietschoten & Houwens bv.
P.O.box 5054, 3008 AB Rotterdam, The Netherlands

ABSTRACT

At the Control Laboratory of the Electrical Engineering Department of Delft University of Technology much research has been carried out during the last 15 years in the field of ship control systems, with the main emphasis on automatic steering of ships. At the moment, the research is concentrated on the design of an autopilot which uses the rudder not only for course keeping but also for roll stabilization, where mostly stabilizing fins have been used up to now. Modeling experiments were carried out and based on their results a simple model of a ship describing the transfer between the rudder and the yaw motions was obtained. The model was subsequently used to design a **R**udder **R**oll **S**tabilization, **RRS**, autopilot.
The autopilot was extensively tested with two types of digital computer simulations and their results were verified with scale-model experiments in open water and full-scale trials with a ship of the Royal Netherlands Navy.
The experiments demonstrated that an RRS autopilot, designed and tested by means of digital computer simulations, is able to realize roll reduction during full-scale trials comparable to that achieved by computer simulations. This makes it attractive to examine new ship designs with respect to the applicability of an RRS autopilot by means of computer simulations. This paper reports the results of these experiments.

1. INTRODUCTION

At the Control Laboratory of Delft University the computer aided design package PSI has been developed. This Interactive Simulation Program, **PSI**, can be used for simulation purposes, but due to its optimization possibilities it can also successfully be applied to the design of a controller for linear as well as for non-linear systems. The combination of a powerful simulation package with the optimization facility makes PSI especially attractive for control systems which cannot easily be designed with analytical methods as well as for modeling based on measurement results.

At the Netherlands **S**hip **M**odel **B**asin (**NSMB**) a maneuvering simulator has been developed, in which more extensive models based on a hydrodynamical approach have been implemented. These models have not only been used to test the controllers developed, but also to serve as a basis for a more simple model.

The present paper describes the digital simulations which were carried out at the Control Laboratory using PSI and at the NSMB using the maneuvering simulator. It shows that their results agree well with those obtained in the full-scale trials with a ship of the Royal Netherlands Navy. This makes it possible, once a hydrodynamic model of a ship is known, to use PSI to investigate whether roll stabilization by means of the rudder is possible for that particular ship. Recently

this idea of stabilizing a ship by means of the rudder has received renewed interest. Searching through the literature, we found publications on this subject written by Cowley and Lambert, 1972, and by Cowley and Lambert, Carley, Lloyd, 1975. In 1980, Baitis reported succesful trials with an RRS-system combined with manual control of the heading. In a paper of Kallstrom (1981) the idea of RRS was also mentioned. In 1981 Van Amerongen and Van Cappelle described a simple mathematical model which can be used as a basis for the design of an RRS-controller. Trials with an RRS-controller based on this model are described in Van Amerongen, Van der Klugt and Van Nauta Lemke, 1983, and in Van Amerongen, Van der Klugt and Pieffers, 1984.
The research on the RRS project is being carried out at Delft University in close cooperation with the company Van Rietschoten & Houwens and with the Royal Netherlands Navy.

The paper is organized as follows: Section 2 describes the mathematical models which are used for both controller design and simulation and it describes the controller as well. Section 3 gives a short introduction to the interactive simulation program PSI and its application to the design of the controller and a state estimator. Section 4 shows some results of the various types of simulation which were carried out and gives a comparison to full-scale experiments, while Section 5 shows how PSI is used to investigate whether RRS is useful when applied to a new ship design. Finally, Section 6 summarizes the conclusions.

2. MATHEMATICAL MODELING

The motions of a ship in waves depend on

- the dynamics of the ship,
- the disturbances, and
- the controller output, which is influenced by the steering machine.

The ship's dynamics

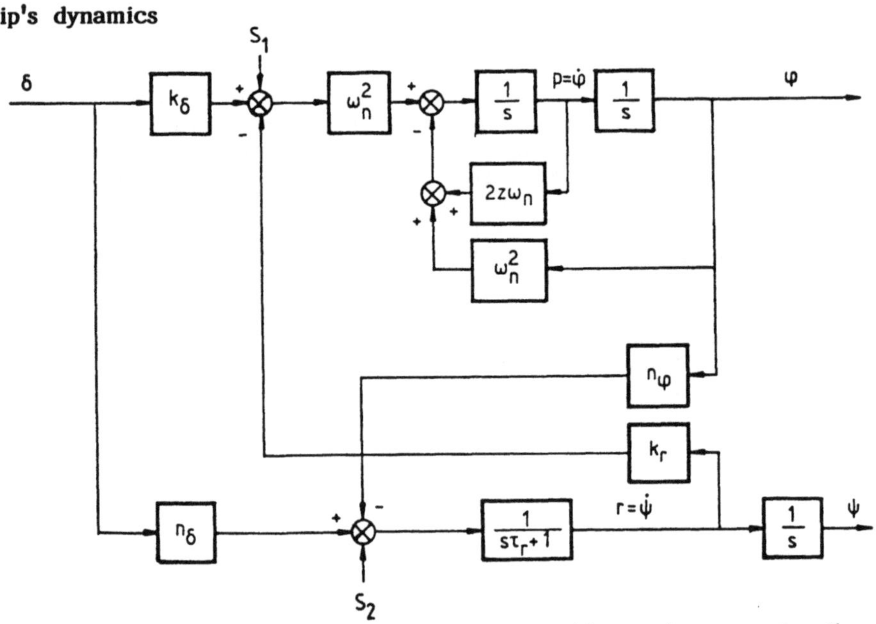

Fig. 1 Simplified dynamics between rudder and yaw and roll.

The basic equations which describe the motions of a ship can be derived from hydrodynamics. Full-scale trials (See van Amerongen and Van Cappelle , 1981) indicate that a good description of these motions is obtained by the model of figure 1.

In this figure δ is the rudder angle,
ϕ is the roll angle,
and ψ is the heading angle.

It should be noted that the transfer from rudder to roll has a non-minimum phase character. A typical response is given in figure 2. For reduction of the roll angle only high-frequency rudder motions can be applied. Low-frequency rudder motions not only cause yaw, but also have the opposite effect on roll than high-frequency rudder motions do. This can be clearly seen in figure 2.

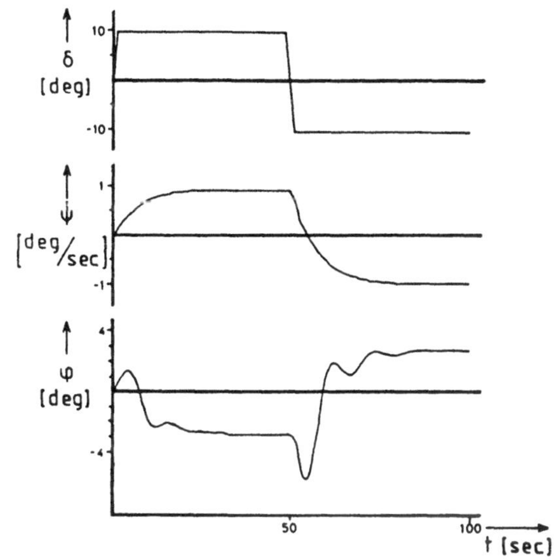

Fig. 2 Response of the rudder-roll transfer

The disturbances

Waves are the most important disturbance with respect to roll. They can be described by means of a frequency spectrum, for instance the Bretschneider spectrum, recommended by the 12th ITTC (See, for instance, Bhattacharyya, 1978):

$$S(\omega) = \frac{691}{\bar{T}^4 . \omega^5} \left(\frac{H_{1/3}}{2}\right)^2 . \exp \frac{-691}{\bar{T}^4 . \omega^4} \quad m^2 s \qquad (1)$$

where $\bar{T}$ is the average period

and $H_{1/3}$ is the significant height of the waves.

It is possible to obtain a more simple approximation of the wave spectrum by means of white noise and a shaping filter, as is shown in figure 3.

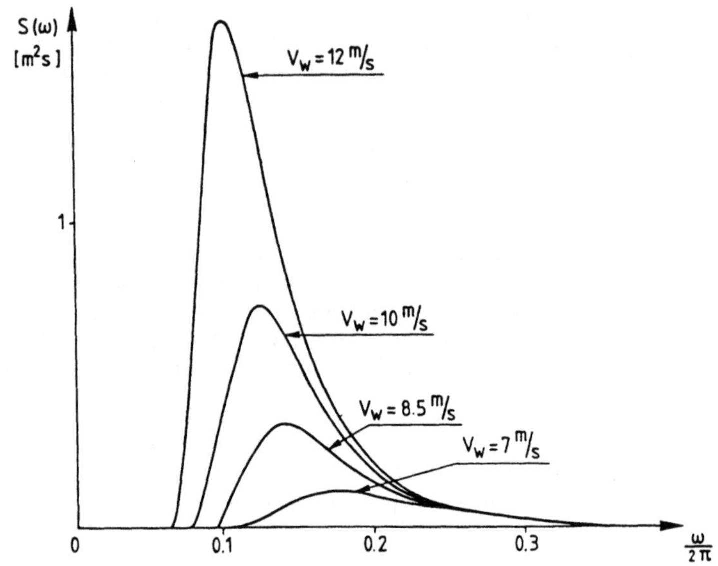

Fig.3 Simulation of the wave spectrum

The disturbance signals which can be computed on the basis of these spectra can be added to the model of figure 1, as indicated in that figure.

The steering machine

For the purpose of the simulations the steering machine is sufficiently accurately described by the block diagram of figure 4. When the rudder is going to be used for reduction of the roll motions it should be able to follow the controller signal without a noticeable phase lag. Van Amerongen and Van Nauta Lemke, 1982, show that this demand puts restrictions on either the maximum rudder angle or the maximum rudder speed. In de controller design care must be taken to ensure that the derivative of the controller output signal is less than the maximum rudder speed in order to prevent phase lag.

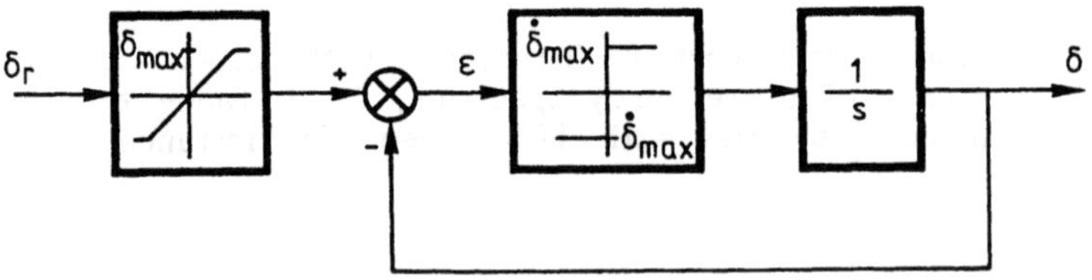

Fig. 4 Model of the steering machine

The controller

In a rudder-roll stabilization system only one input (the rudder) is available to control two outputs (yaw and roll). Decoupling of yaw and roll can be obtained by restricting low-frequency rudder motions to control yaw while high-frequency rudder motions are restricted to roll reduction. The latter is also necessary due to the non-minimum phase character of the roll response. Van Amerongen, 1984, describes an adaptive state estimator which yields low-frequency estimates of the heading and rate-of-turn signals. This estimator, which combines the concepts of model-reference adaptive systems (MRAS) and Kalman filtering, is suitable for application in an RRS autopilot as well.

In general, the roll angle will contain a constant offset which depends on how well the ship has been trimmed and on the disturbing moment of the wind. Because a constant roll angle cannot be reduced by RRS the controller should disregard it. Basically, a simple high-pass filter is suitable for this purpose.

When only measurements of the roll angle itself are available the roll rate, and eventually the roll acceleration, have to be estimated. How this can be done by means of Kalman filtering techniques is described by Van Amerongen, Hoogenraad and Van Nauta Lemke (1984).

The following controller was applied:

$$-\delta_r = \left[K_\phi \cdot \phi + K_{\dot\phi} \cdot \dot\phi + K_\psi \cdot \psi + K_{\dot\psi} \cdot \dot\psi \right] \quad (2)$$

where $\dot\phi$ denotes $d\phi/dt$ and so on.

The influence of the disturbance on the motions of the ship depends not only on the spectra given but also on the angle of incidence of the waves. The latter not only affects the amplitude of the motions but also causes a frequency shift. Therefore, it is essential that the controller parameters be adjusted by an adaptation mechanism. The experiments described in this report used fixed controllers for each ship speed and disturbance, although small variations were made in order to investigate the sensitivity of the system to those variations.

3 APPLICATION OF PSI

At the Control Laboratory of Delft University the computer aided design package PSI has been developed. It is a combination of a powerful simulation package and an optimization facility based on a fast hill-climbing algorithm.
Furthermore, it is possible in PSI to make use of measurement data if they are in a suitable format.
PSI is able to calculate the "best-fit" model of a system by means of simulation and optimization. Linearity assumptions for model or criterion are not necessary. This approach is illustrated in Fig. 5.

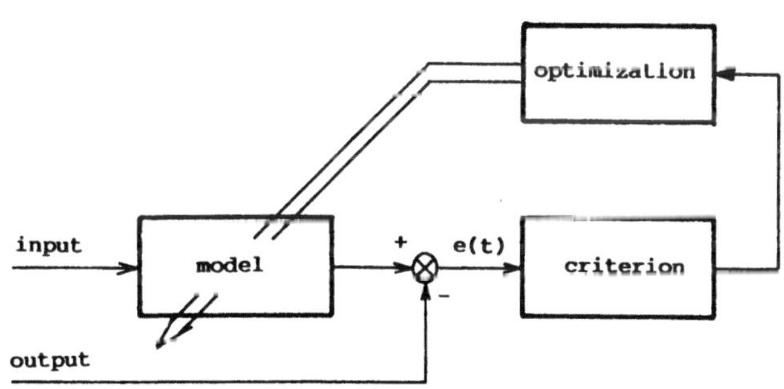

Fig. 5 Identification via simulation and optimization.

A criterion is defined, based on the error between the output of the process and the output of the (adjustable) model. Both are excited by the same input signal. The output of the model is obtained by using simulation techniques. Therefore, the model may be described by continuous parts, discrete parts, non-linear or logical elements or any combination of these. The output of the process and the input signal are obtained from measurement data.
The optimization algorithm is able to find the model parameters of a (non-)linear model with a user-defined structure and a user-defined criterion.

A similar approach can be used to design a controller. This is illustrated in figure 6.

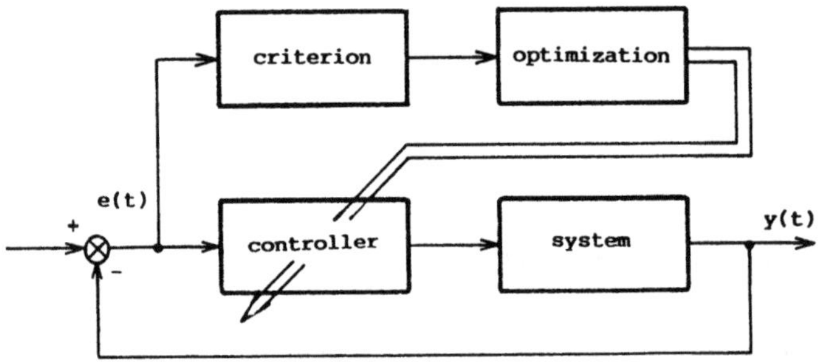

Fig. 6 Designing a controller using simulation and optimization.

Simulation techniques are used to simulate the controller and the process and to calculate the error signal e. A criterion is defined, for instance, based on this error signal and the output of the controller, which can be optimized with respect to the parameters of the controller. So, any (non-)linear system and any controller configuration can be used with any criterion. Finite or infinite-dimensional constraints can be included, via penalty functions, in the criterion. Even the combination of a discrete controller which controls a (non-)linear continuous system offers no problems.

Using PSI, the controller has been designed based on the following criterion:

$$J = 2.\max |\phi| + 5.\max |\psi| \qquad \text{for } 0 < t < T$$

where max $|\phi|$ and max $|\psi|$ are the maximum values of the roll and heading signals during the observed time interval T.

4 RESULTS OF SIMULATION AND FULL-SCALE TRIALS

The first experiments were carried out with the interactive design package PSI (Van Amerongen and Van Nauta Lemke, 1982). Figure 7 shows a typical example of such a PSI simulation. Two equivalent ship models are responding to the same disturbances. The solid lines show the performance of the ship with an RRS-controller, while the dotted lines show the performance of the ship with only a course controller. In this particular run a reasonable roll reduction was obtained.

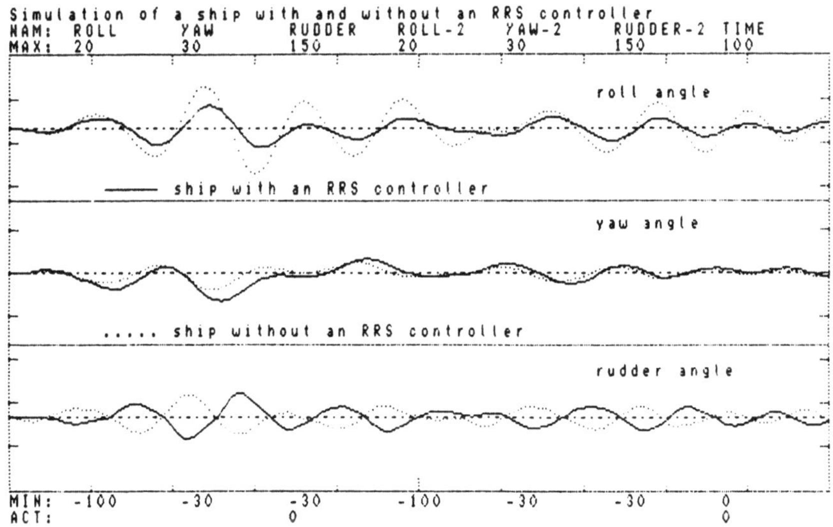

Fig. 7 Example of a PSI simulation

Afterwards, the controller was implemented in a digital computer (DECLAB 11/03). Initial experiments were carried out by testing the controller in combination with a simple analog simulation based on the models of the ship, steering machine and disturbances described in Section 2. After the successful completion of these experiments, a series of tests was carried out by coupling the RRS autopilot with the computer of the Netherlands Ship Model Basin's (NSMB) maneuvering simulator, which contained a more extensive model based on a hydrodynamic approach (Van Amerongen and Van der Klugt, 1983).

No knowledge of the model implemented at the NSMB-computer was used for the controller design, except that it was known to be a model of a type similar to the ship. This approach more or less approximates a realistic situation. It had to be demonstrated that the controller settings computed with the simple model were also able to control the more complex mathematical model at the NSMB.

The following parameters were varied:

The angle of incidence of the waves (γ), defined in figure 8.

The values of γ which were selected were

γ = 60 degrees
γ = 90 degrees
γ = 120 degrees

The first run of each series was made with the RRS controller switched off. This yields a reference for the other runs. Three different rudder speeds and maximum rudder angles were selected:

$\dot{\delta}_{max}$ - δ_{max} = 6 deg/s. - 10 degrees
$\dot{\delta}_{max}$ - δ_{max} = 15 deg/s. - 20 degrees
$\dot{\delta}_{max}$ - δ_{max} = 20 deg/s. - 30 degrees

In each situation the controller gains were selected as optimally computed before (nominal), 30 percent higher (+ nominal) and 30 percent lower (- nominal) in order to investigate the sensitivity to variations in the controller gains. The signals ϕ, ψ and $\dot{\psi}$ were obtained by means of the adaptive state estimator with the noise-suppressing properties described in Section 2. Most runs were made with sea state 5, and a few with sea state 6.

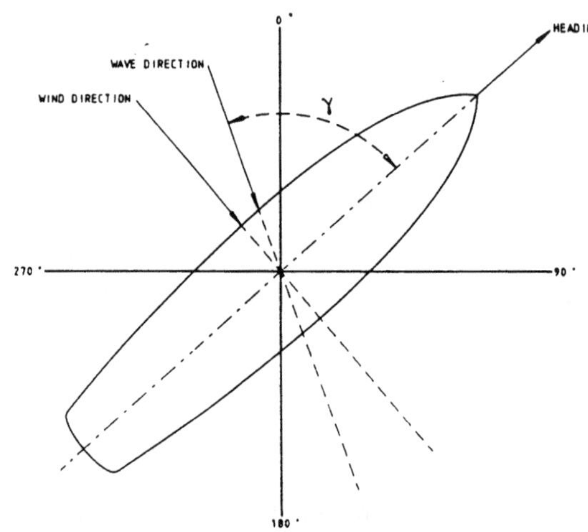

Fig. 8 Definition of the angle of incidence.

The following criterion was used to compare the various runs:

$$J = \left[1 - \frac{\sigma^2_{\phi}\text{, closed}}{\sigma^2_{\phi}\text{, open}}\right] \times 100 \qquad (4)$$

where $\sigma^2_{\phi} = \overline{(\phi - \bar{\phi})^2}$

and ϕ_{open} is the roll angle without an RRS-controller and
ϕ_{closed} is the roll angle with an RRS-controller.

The results are summarized in figure 9, which gives the reductions for various angles of incidence as a function of the maximum rudder speed. All runs were successful except for those with a rudder speed of 20 deg/s, where the controller gains were apparently too high.

The experiments indicate that the rudder speed in particular is an important parameter which determines the maximally achievable roll reduction. Rudder speeds which are common at present (3 to 7 deg/s) are generally too slow. A rudder speed of 15 deg/s appears to yield a considerable improvement. The variations made in the controller parameters indicate that the system is rather sensitive to these parameters. Settings which are too low are preferable to those which are too high, which may cause instability.

The next series of trials were carried out with an 8 meter-long radiographically controlled scale model of the same naval ship used during the simulations (Van Amerongen, Van der Klugt and Pieffers, 1984). These trials, carried out at the Haringvliet (a former sea arm in the South West of the Netherlands) led to the development of an important adaptive feature of the autopilot, the Automatic Gain Control (AGC) system, for which a patent is pending. As soon as the AGC detects that the demanded rudder speed exceeds the limitations of the steering machine, the controller is automatically adjusted. This reduces for some time the roll-reduction capability of the system as well, but it prevents the system from becoming unstable and it considerably improves the course-keeping performance.

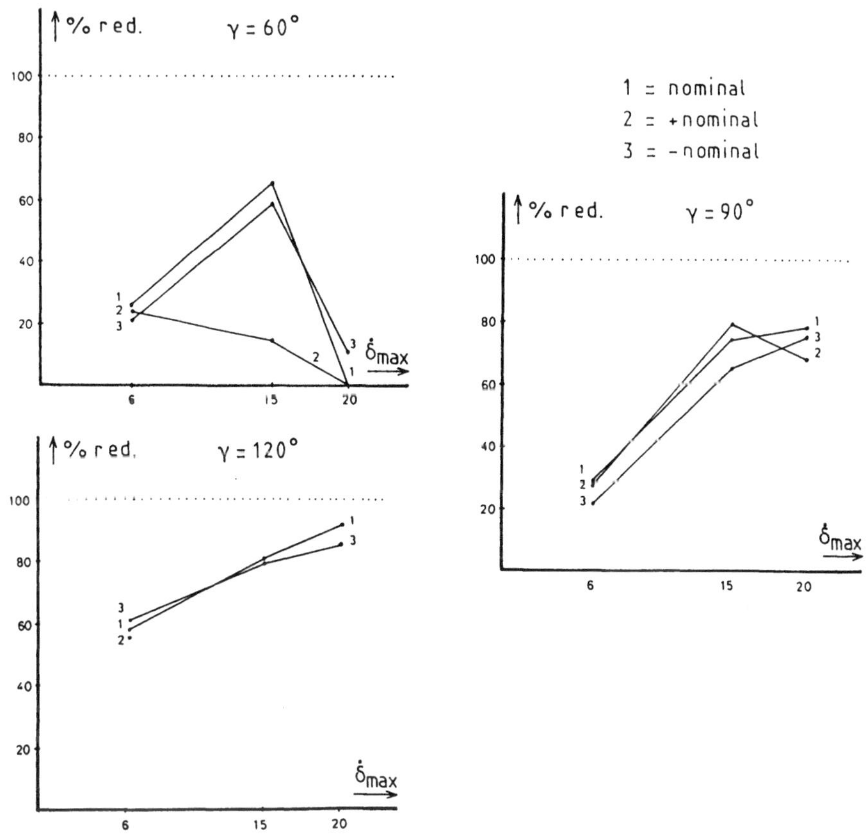

Fig. 9 Comparison of the controller performance with different angles of incidence and different rudder speeds.

The next experiments were carried out on board a ship of the Royal Netherlands Navy, similar to the ship whose dynamics were simulated in the simulation experiments. The experiments were carried out on the North Sea and the circumstances appeared to be almost ideal for this kind of trial; wind forces up to Beaufort 10 (Van Amerongen, Van der Klugt and Pieffers, 1984). The main aim of these full-scale trials was to verify the earlier simulation results on a real ship and to test the AGC-system. Figure 10 gives a typical example of such an experiment. This figure shows a comparison of the RRS-system with ASA, adaptive ship's autopilot, earlier developed at the Control Laboratory (Van Amerongen, 1982, 1984), and with the ship's autopilot. In this figure the roll reduction (64%) is most clearly visible. The improved course-keeping performance of the RRS-controller can also clearly be seen, especially in comparison with the ship's autopilot.

The results from the full-scale trials were then compared with the simulation results obtained at the NSMB, not only to validate the latter but also to extrapolate the simulation results for higher rudder speeds of the actual ship. In figure 11 the results of the experiments at the NSMB are summarized.

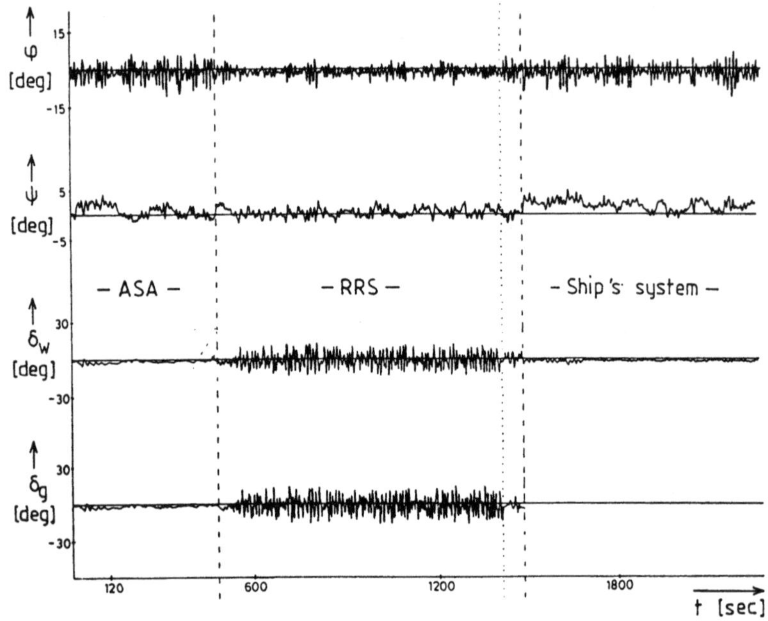

Fig. 10 A qualitative comparison between the ASA autopilot, an RRS controller and the ship's autopilot

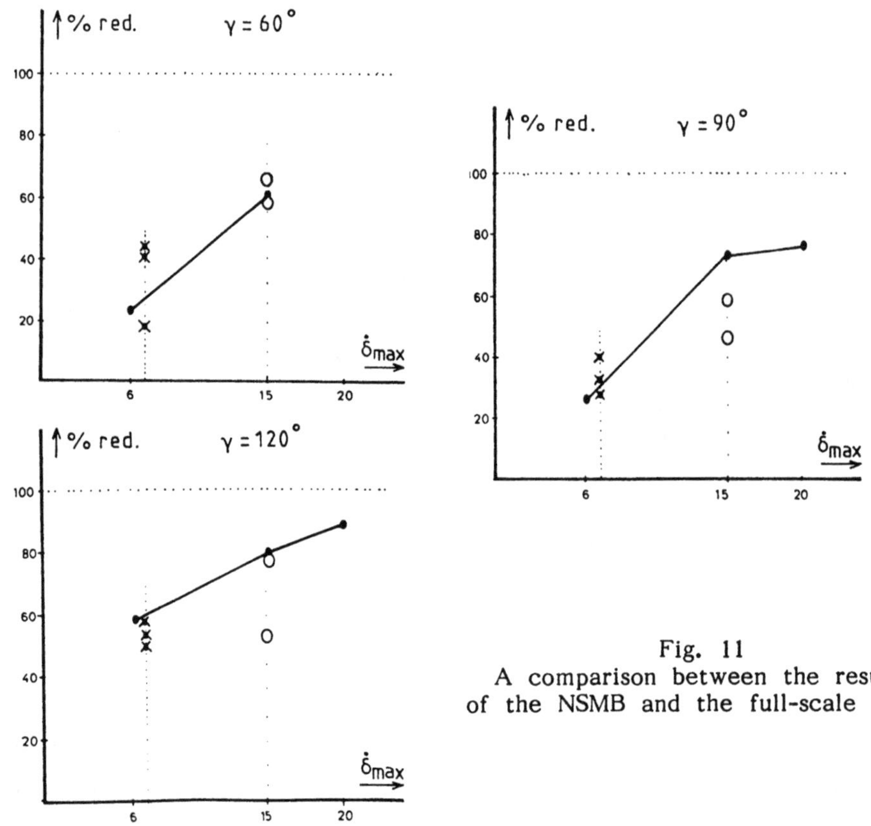

Fig. 11
A comparison between the results of the NSMB and the full-scale trials.

In this figure a series of reductions obtained during the full-scale trials, with a rudder speed of 7 deg/s has been added (plotted as X). Because it was not always possible to determine a unique wave direction, the frequency spectrum at sea was less sharp than during simulation. Furthermore, the angles of incidence of the waves do not completely match.
When these restrictions are kept in mind it seems allowable to make the comparison between the full-scale trials and the simulation experiments. It can be observed that the reductions obtained at full scale are indeed comparable to those of the simulation.

In order to compare the present fin-stabilizer system to the reductions which may be anticipated for a maximum rudder speed of 15 deg/s, the reductions of the former have been added in figure 10 as well. They are plotted as open circles. It can be observed that the results of the present fin-stabilizer system are comparable to the simulation results of an RRS-system with a maximum rudder speed of about 10 to 15 deg/s. It may thus be expected that the roll reduction with RRS and a rudder of 15 deg/s will be at least as good as with present fin stabilizers.

5 APPLICATION OF PSI TO IMPROVE THE DESIGN OF SHIPS

The above-mentioned full-scale trials were carried out on a ship which is not readily available for RRS trials. Therefore, an alternative was considered desirable. A likely candidate is the 'Zeefakkel', the training vessel of the Royal Netherlands Naval College. In order to investigate whether the Zeefakkel is indeed suitable for RRS trials, some modeling trials were performed. The measurement data was transformed to a suitable format which allowed modeling at the Control Laboratory.
At that time PSI was not as powerful for identifying processes as IDENT, a special identification program (Van Amerongen and Van Cappelle, 1981). Therefore, it was decided to use IDENT to obtain the desired model parameters, based on the model of figure 1. Figure 12 shows a typical identification run.

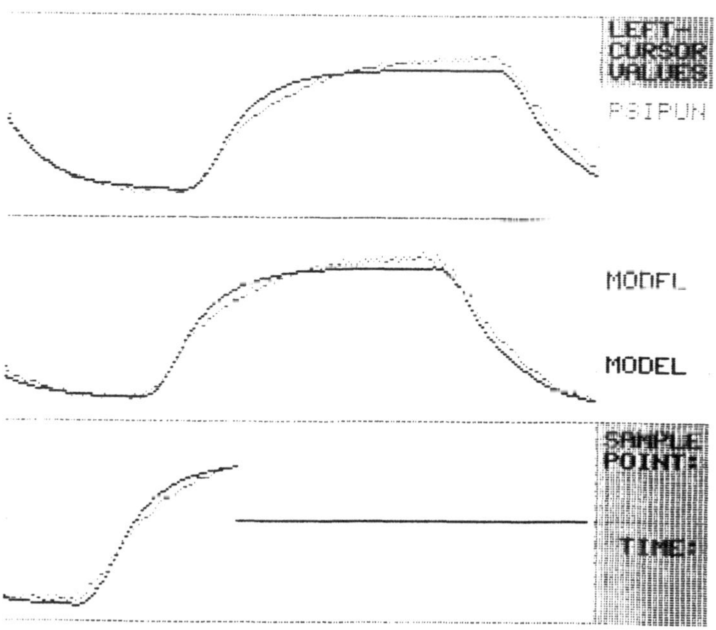

Fig. 12
Identification of a model of the 'Zeefakkel'

The identification was performed in two steps. First the rudder-to-yaw transfer was identified, then the performance from rudder-to-roll was identified.
These steps were repeated several times to identify the influence of the roll on the yaw as well. Figure 12a shows the identification of the rudder-to-yaw transfer of one particular run while figure 12b shows the identification of the rudder-to-roll transfer of the same run. The dotted lines give the measurement data of the above mentioned zigzag trials, while the solid lines give the identification results. As this figure indicates a close fit was possible.

After these identifications, experiments were carried out where PSI was used to investigate whether it is possible to obtain a reasonable roll reduction with this particular ship. Figure 13 shows a particular simulation run. The dotted lines represent the performance of the ship without the RRS controller while the solid lines present the performance of the same ship with an RRS controller. The results revealed that, even if the 'Zeefakkel' were to be equipped with a fast (15 deg/s) rudder, the amount of roll reduction that could be obtained is less than that possible at the above-mentioned naval ship.

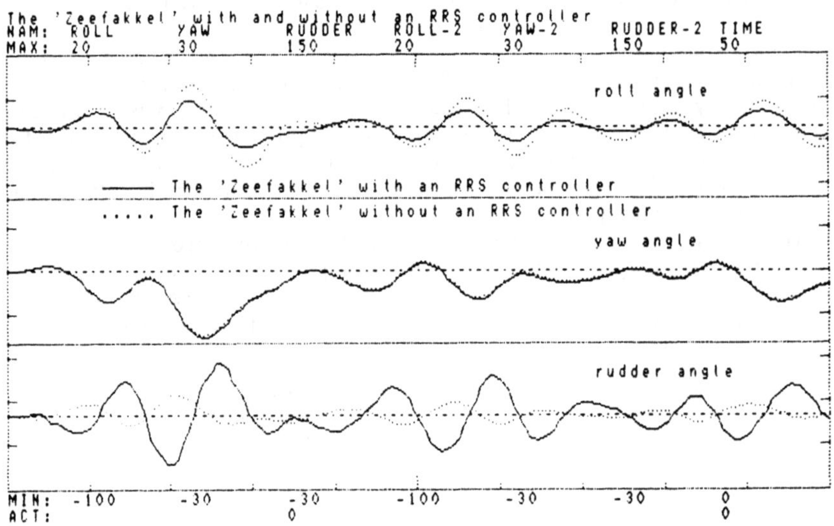

Fig. 13 Simulation of the 'Zeefakkel' with and without an RRS-system

Recently, a project has been started in cooperation with the Royal Netherlands Navy and the NSMB dealing with designing a new ship, equipped with an RRS-system. An extensive model of this new design has been derived based on a similar hydrodynamic approach, as mentioned in Section 3.
Zigzag trials were carried out with this model and the data were transformed to a suitable format which allows modeling at the Control Laboratory. PSI will be used to verify whether a reasonable roll reduction can be obtained with this particular ship. This way PSI offers an economically attractive opportunity to investigate the influence of changes of the original ship design on the maneuvering characteristics.

6 CONCLUSIONS AND SUGGESTIONS

In this report the results of a series of trials have been represented. Modeling trials have been carried out which resulted in a simple model of a naval ship. A RRS controller has been designed be means of the simulation package PSI and subsequently the following experiments have been carried out:

1 The controller was tested in a digital simulation where the controller and simple model of the ship, steering machine and disturbances were implemented in PSI.
2 The controller was tested in combination with an analog simulation of the same models while the controller was implemented in a digital computer.
3 The controller was tested on a more extensive model of the same ship based on a hydrodynamic approach.
4 The controller was tested on a scale model of the same ship.
5 The results of these experiments were verified during full-scale trials.

Most of these experiments agreed well. This indicates that in order to test the applicability of an RRS system, it is allowed to use the results of one type of experiments to predict the results of an other type of experiments. This knowledge has been used to predict whether an existing ship (the 'Zeefakkel') and a new ship design were suitable to be equipped with an RRS system and it has been used to predict the performance of a ship with an RRS system and with fin stabilizers.
The simulation package PSI appears to be a powerful aid in investigating the maneuvering characteristics of ships.

In the near future PSI will be extended with the possibility to identify system models using measurement signals consisting of 1000 points. That would make PSI even more suited to identify the non-linear aspects of the maneuvering characteristics of a ship.

REFERENCES

J. van Amerongen and J.C.L. van Cappelle, "Mathematical modeling for rudder roll stabilization", 6th Ship Control Systems Symposium, Ottawa, Canada, 1981.

J. van Amerongen and H.R. van Nauta Lemke, "Rudder Roll Stabilization", Proceedings of the 4th ISSOA International Symposium on Ship Operation Automation, Genua, Italy, pp. 43-50, September 20-22, 1982.

J. van Amerongen, P.G.M. van der Klugt and H.R. van Nauta Lemke, "Roll stabilization of ships by means of the rudder", Proceedings Third Yale Workshop on Applications of Adaptive Systems Theory, New Haven, Conn., USA, pp. 19-26, 1983.

J. van Amerongen, "Adaptive steering of ships - a model reference approach", Automatica, vol. 20, no. 1, 1984, pp. 3-14.

J. van Amerongen, P. Hoogenraad and H.R. van Nauta Lemke, "Estimating the waves spectrum and the roll angle and its derivatives with an extended Kalman filter", 7th Ship Control Systems Symposium, Bath, U.K., 1984.

A.E. Baitis, "The development and evaluation of a rudder roll stabilisation system for the WHEC Hamilton Class", DTNSRDC Report, Bethesda, Md. USA, 1980.

R. Bhattacharyya, "Dynamics of marine vehicles", Jonh Wiley & Sons, New York, 1978.

P.P.J. van den Bosch, "Software tool for control system sesign", Journal A, vol. 22, no. 2, 1981, p. 55-62.

P.P.J. van den Bosch and J. van Amerongen, "Software for Simulation and Optimisation Applied to Ship Control Systems", 7th Ship Control Systems Symposium, Bath, U.K., 1984.

J.B. Carley, "Feasibility study of steering and stabilising by rudder", Proceedings 4th Ship Control Systems Symposium, The Hague, The Netherlands, 1975.

W.E. Cowley and T.H. Lambert, "The use of the rudder as a roll stabiliser", Proceedings 3rd Ship Control Systems Symposium, Bath, UK, 1972.

W.E.Cowley and T.H. Lambert, "Sea trials on a roll stabiliser using the ship's rudder", Proceedings 4th Ship Control Systems Symposium, The Hague, The Netherlands, 1975.

C.G. Kallstrom, "Control of yaw and roll by a rudder / Fin stabilisation System", 6th Ship Control Systems Symposium, Ottawa, Canada, 1981.

A.E.J.M. Lloyd, "Roll stabilisation by rudder", Proceedings 4th Ship Control Systems Symposium, The Hague, The Netherlands, 1975.

A Modular Manoeuvring Mathematical Model for Warship Simulation

IR McCALLUM
University of Wales Institute of Science & Technology and
Maritime Dynamics Ltd

R MATTHEWS
Maritime Dynamics Ltd

Summary

The changing pattern of usage of ship simulators over the past few years is examined, and the trend towards individual projects noted. The need for more flexible ship manoeuvring mathematical models is then discussed, and the developemnt of a particular model, representing two warship types, is briefly described. Towing tank tests were used to augment the limited full scale results available. Validation of the model is discussed, and a proposal for international definition of agreed mathematical model sub-system interfaces made.

Introduction

Ship simulators have now been in regular use in most maritime countries for about eight years. The major development during that time has been the change in the uses to which simulators have been put. At the end of the last decade, simulator studies were concerned largely with the open water manoeuvring of large ships, typified by the early studies into the manoeuvring of tankers into Valdez, and training was largely concerned wih the then new topic of bridge procedure training.

These activities do not place any great constraints on the ship manoeuvring mathematical model, as it is comparatively simple to produce a model which will perform in deep water, with the ship going ahead at cruising speed. Mathematical models of a range of ships were produced for simulators, generally of the hydrodynamic derivative type, (1), (2), which, while costly and time consuming to produce, nevertheless enabled simulators to fulfil their functions adequately. It is interesting to note that it is only in the past two years that there has been any serious attempt towards categorising the mathematical models used in simulators.

The main trends in ship simulator use in the past few years have been away from the study of large ships in deep water. It was found early on that there are few basic maneouvring problems with very large ships in open water. Simiilarly, the market for bridge procedure courses has diminished, both because of the general maritime recession, and also because there is now a large body of professional mariners who have done the standard courses.

What is now attracting the attention of the simulator operators is the study of problems associated with close quarters manoeuvring of ships of all sizes. New ferries are being built which are larger than previous ones, but still have to berth in the same jetties, offshore supply vessels have to manoeuvre very close to rigs in all weathers, trials are being carried out to determine whether Panamax ships can pass safely in the Panama canal, and very many new types of ship are entering service, for which the operators need to determine safe operating methodologies, in both normal and failure regimes.

Training courses now tend more to be centred around a specific ship or port related problem, as opposed to procedural methods. Simulator users want their particular ship and their port to be simulated for a particular training course or investigation. This emphasis on the particular, rather than the general, has caused the role of the mathematical model to be re-examined. For many tasks, it is necessary to produce a special mathematical model which may readily be inserted into the simulator, and whose structure is specially tailored for the task in hand. There are now several types of model available for simulators, which are able to represent the behaviour of ships in all manoeuvring regimes, and which may be produced comparatively quickly and cheaply. It remains of course the responsibility of the simulator operator to determins which of the available models is adequate for his task, bearing in mind the constraints of accuracy, comprehensiveness, price and time available to produce the model.

The first comprehensive CGI based ship simulator to have installed a suite of several different types of mathematical model was CASSIM. It was found, immediately after it was opened in 1982, that the vast majority of tasks which the simulator was asked to carry out involved the production of a special mathematical model. Frequently, there was only a small allowance of time and cost to provide the mathematical model, and so it became necessary to devise a suite of programs, using an auxiliary motion computer, which enabled any one of three widely different types of mathematical model to be used, (3). The overall concept was that the most suitable type of mathematical model for a particular task would be selected by the Project Officer, and a model made using any of the three different forms of analysis available. The model could then be "slotted in" to the simulator.

The aim of this paper is to outline the methodology whereby this type of modular model may be used, to describe the production of a particular type of model for a new type of ship, and to propose some standards to be used for the production of a range of simulator mathematical models.

The Modular Model

The basic concept of the Modular Model is very simple. It consists of no more than dividing the model structure into a number of identifiable sub-systems, such as rudder control, propeller, hull forces, etc. The inputs and outputs of these sub-systems are then defined in a rigorous manner, so that different models of each sub-system can be used, depending on the requirement of a particular task.

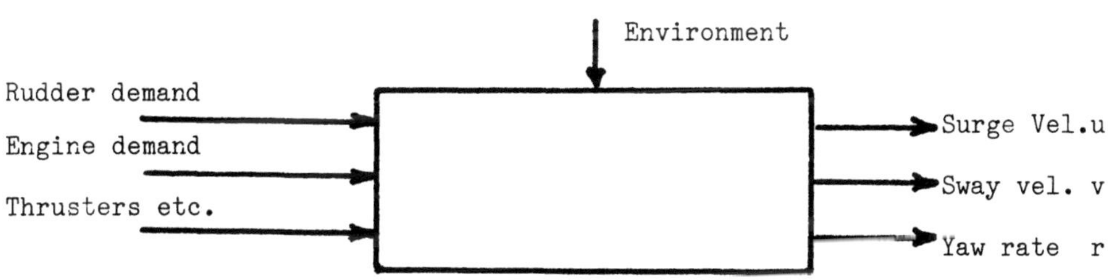

Fig. 1. Basic mathematical model structure

Fig 1 indicates the overall requirement of a simulator mathematical model, which is to receive inputs of the effectors and environment, and to output the ship motions. Fig 2 shows one possible division of a mathematical model into sub-systems, but there are many different ways in which the model may be so divided.

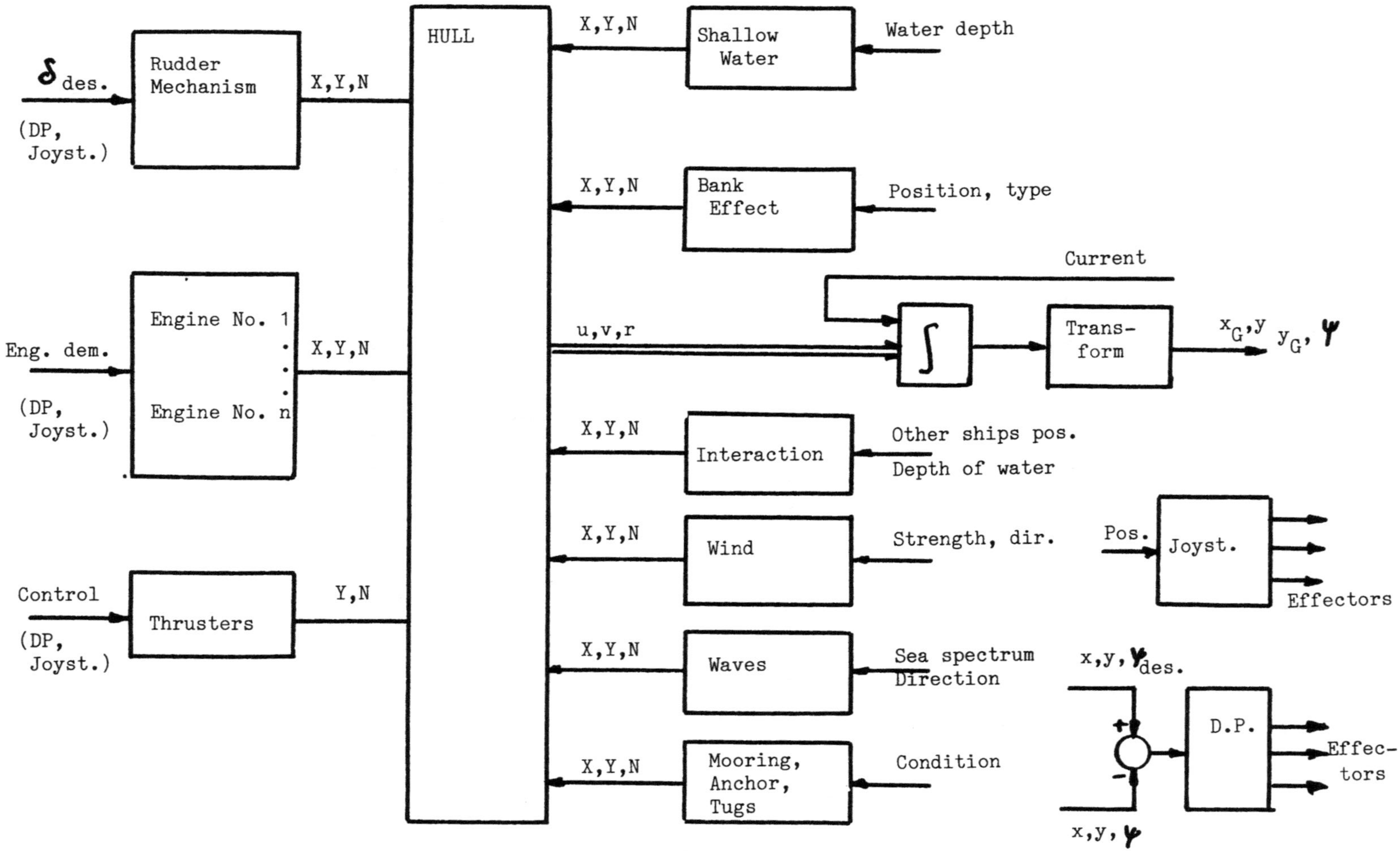

Fig. 2. Modular Model - Possible Structure.

It is obviously important that the divisions represent, as far as possible, identifiable physical aspects of the ship. Thus, the rudder control and engine control mechanisms may be easily identified, as they have clear inputs and outputs in physical terms. The identification of such systems as hull forces is more difficult conceptually, as it is not easy to separate the contributions of the hull, rudder and skeg in determining the overall forces acting on the ship.

The treatment of environmental effects is also problematical. Conventionally, Shallow Water Effect is implemented by modifying the coefficients or derivatives of the basic hull model, and it is reasonable that this should be so, as the hull forces are themselves affected by the presence of the seabed. However, the conventional method of simulating Bank Effects is to add separate bank forces onto the model, using the assumption that the presence of banks causes additional forces to act on the hull. It is not easy to distinguish conceptually between the effect of shallow water and that of a bank.

A major advantage of the Modular Model is that the complexity of the model, and hence its cost, can be closely related to the overall need for the simulation. Thus, a simulator study on the ability of pilots to use a new type of steering stand for effective track-keeping may have a very sophisticated model of the rudder and steering engine, but is unlikely to need a complex model of the hull hydrodynamics, as the ship will not be used in the more complex manoeuvring regimes. Similarly, a model used for human factors bridge studies may be able to be provided very cheaply indeed, as the behaviour of the ship may not be very critical to the study in hand.

A well designed Modular Model will enable additions to be made to it with the minimum of disturbance. Thus, for example, the addition of a Dynamic Positioning System should be able to be made to the model in much the same way as a real DP system will be interfaced to the ship.

Modelling Warships

A requirement arose during 1984 to produce a mathematical model of two warship types, a Frigate and a Fast Attack Craft. These models were for an advanced radar simulator, but one which had the potential for incorporating a visual system, so the performance of the models had to be to a high specification. The models had to operate in all manoeuvring regimes, and it was important that the basic model structure was able to represent a large range of warship types. At the start of the project, it was not known whether the model structure would end up as being particular to warship types, or whether the methodology would be applicable to other, merchant, hullforms.

Warships manoeuvre in fundamentally different ways from most merchant ships. Their behaviour is governed by the shape of the hull and the position, number, size and type of their control surfaces and effectors. The main differences are caused by their large length-beam ratio, low block coefficient and large power. Rudders are set well away from the hull and in the slipstream of the propellers.

The high speed attainable with the large power available means that the manoeuvring characteristics are speed-dependent, in a way that most merchant ships are not. Turning circles are in general larger at high speeds. Surprisingly, most warships are not intrinsically very manoeuvrable, achieving a turn diameter of about four shiplengths at full speed and full rudder, compared with under two for many merchant ships. A feature of most warships is that their behaviour in turn is typically very stable, with very small pullout angles.

A major problem in producing a mathematical model of a warship is that the details required of their dynamic behaviour are precisely those which the naval authorities are unwilling to release. To resource these models, alternative sources of data were used. These included using a scale model of the required hullform in a towing tank in two modes; As a captive model, to investigate the forces and moments acting on the hull throughout a range of drift angles from zero to 180 degrees, and as a free running model, to investigate the turning behaviour free from the constraints of the towing carriage. In the latter condition, the model was being used as a small ship, and, in order to eliminate the scale effects of using such a model, a mathematical model was made of the actual physical model.

The results of the tank and manoevring basin trials gave a clear indication of the forces and moments on a slender warship hull at angles of attack up to 180 degrees. The basic form of the measured forces and moments were largely as expected:

The lift curve is basically a sine wave, with an argument at twice the drift angle, (Fig 4).

The drag curve is essentially a cosine function, offset to allow for the drag at zero angle of incidence, (Fig 4).

The hull moment is essentially another sine curve, of the same basic shape as the lift curve, (Fig. 3).

An important effect is noticed at small angles of incidence of the hull, in that the slope of the lift force curve is very much less, by a factor of about four, than the value expected from a basic calculation. The diminution of the hull lift in these conditions appears to be a feature of the low aspect ratio of the hull, (4).

When the hull forces and moments are resolved into body centred axes, to produce forces in surge and sway, the effective point of application, the Centre of Pressure, may be evaluated, (Fig.5). This, it is found, migrates from a position well forward at low angles of incidence, to a position abaft the CG at angles over about 100 degrees. The lack of symmetry about the CG is thought to be caused by the smaller longitudinal area of the afterpart of the submerged hull.

A major dichotomy emerges when the results of the tank trials are examined, in that the size of the moment curve of the hull is very much greater than the counter moment exerted by the rudder, at all angles above about 10 degrees of drift angle. In other words the basic hull shape is very unstable directionally. This is in direct opposition to the common manoeuvring experience, in which the ship is very stable directionally under all operational conditions.

Two factors are thought to be relevant:

* At small angles of incidence, (the course keeping condition,) we have seen that the hydrodynamic lift is much reduced, and the resultant drift angle remains at values such that the rudder can effectively counter the tendency of the hull to yaw.
* At larger drift angles, the ship must be moving on a curved path,

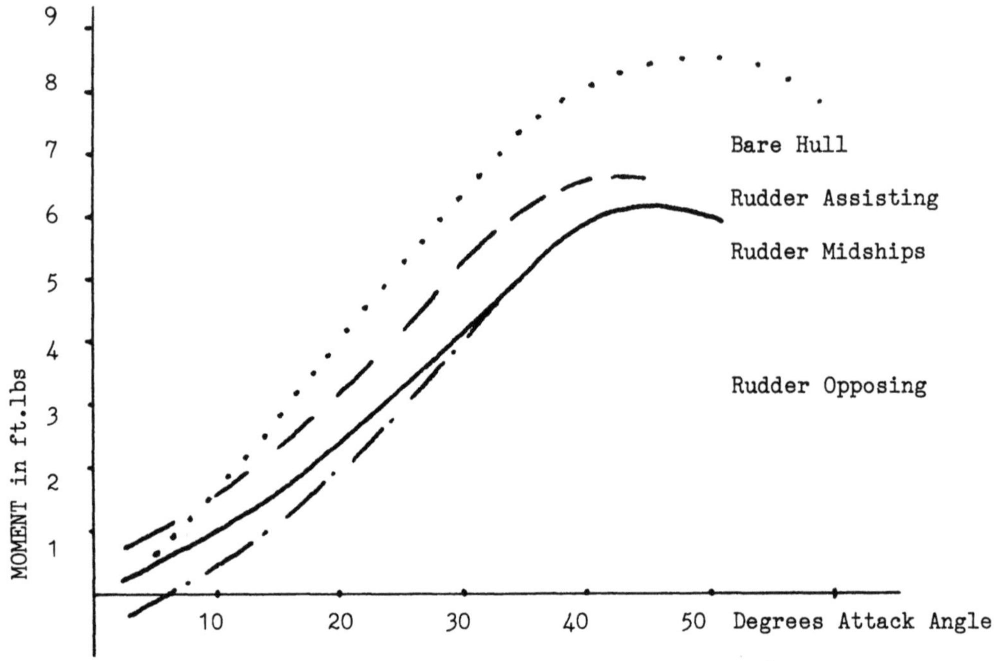

Fig.3. Moment of Frigate Tank Model @ 0.5m/s

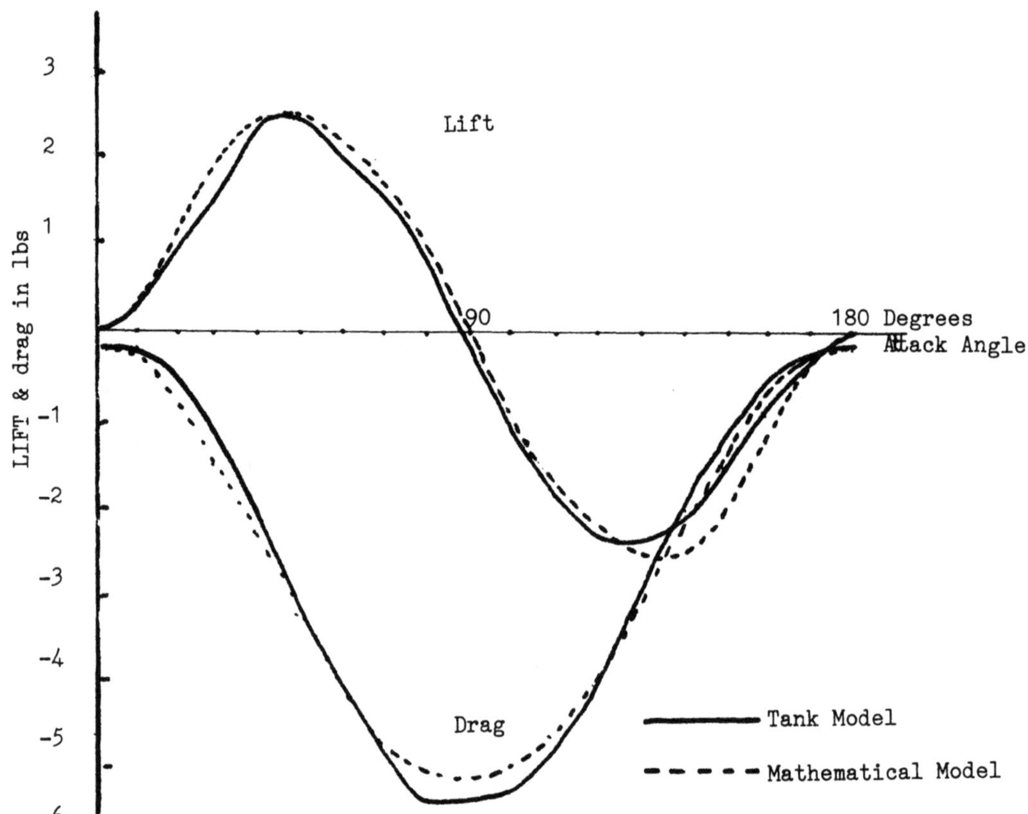

Fig.4. Lift & Drag of Frigate Tank & Mathematical Model

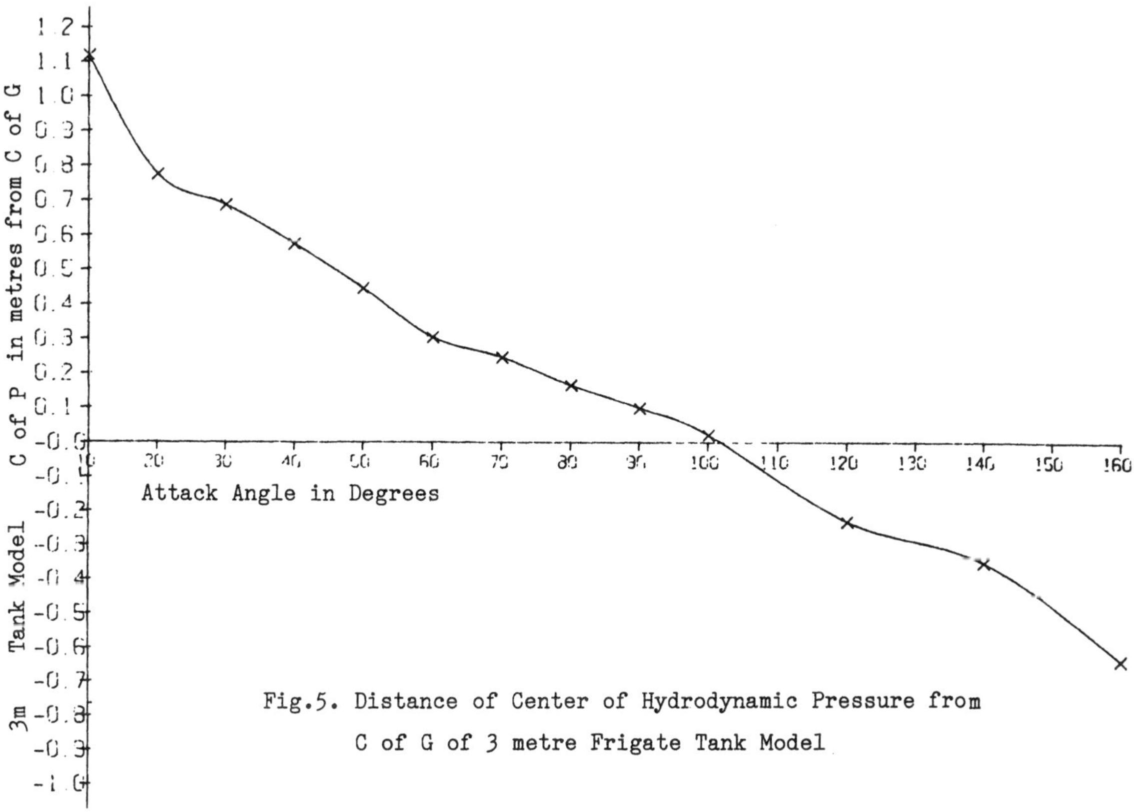

Fig.5. Distance of Center of Hydrodynamic Pressure from C of G of 3 metre Frigate Tank Model

Fig. 6. **Variation of Drift Angle along Length of Turning Ship**

Drift Angle at Bow = 0°
Drift Angle at CG = 10°
Drift Angle at Stern = 25°

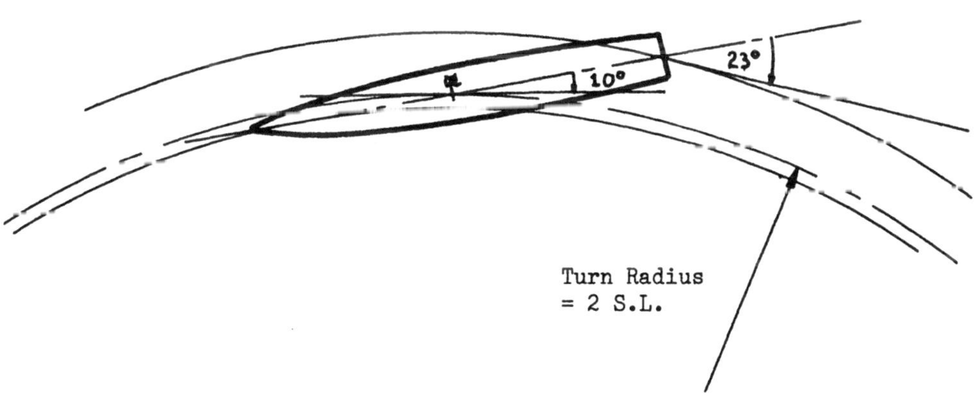

and so the streamlines, instead of being straight, are curved, as shown in Fig. 6. This will have two effects: The overall lift is likely to be reduced, as the angle of the incoming water approaches zero, and the centre of pressure will shift aft. Both these effects will have a pronounced stabilising effect on the hull.

Modelling Methodology

The basic methodology used was to analyse the forces and moments acting on the hull in a direct manner, as described in references (3) and (5). In earlier models the hull and rudder forces were evaluated as functions of the effective drift angle at the hull and rudder, using empirical methods. The results of the tank trials enabled the forces and moments to be modelled more directly. There is however a similarity between the present results and the empirical results of the earlier models.

The basic model structure was already in an elementary type of modular format, in that the rudder and engine controls were in separate parts of the model program, and the rudder forces are related to the local streamlines, again in a seperate part of the program. No attempt has yet been made however to separate out the various elements of the program in a formal manner. If the concept of modularity becomes at all widely accepted, this will have to be done.

Model Validation

The validation of mathematical models is customarily carried out in three phases:

Verification: The model structure is examined for soundness, and a number of elementary tests carried out to ensure that, for example, the ship turns to port when port rudder is applied. For the present model, verification will include establishing that the forces simulated represent those found in the tank to an acceptable degree of accuracy.

Quantitative Validation: This consists of running the model, over as wide an operating range as is possible, (or desired), and ensuring that it performs in the same way as the ship throughout that range. It is of course essential to use independent data to validate than that which was used to derive the model in the first place. For the warships, this phase is not as thorough as might be wished in the almost total lack of adequate ship manoeuvring data. Some reliance was therefore placed on the free running model results.

Qualitative Validation: It is difficult in nearly every case to acquire sufficient full scale trials data to ensure that the model performs well in every manoeuvring regime, and so a form of validation frequently used is to invite experienced mariners to operate the model in real time in a simulator and comment on its behaviour. This method of validation is at best unreliable, but the important area of transient turning behaviour might be able to be checked better this way than by examining plot data.

Validation plots for the warship models are shown in Figs 4 and 7. In Fig 4, the results of modelling the lift and drag curves of the tank trials are shown. It is a matter for some conjecture whether there is any requirement for an exact fit here, as the precise shape of the curves is likely to depend on the operational conditions. Fig 7 shows a comparison between simulated results and actual ship trials for a frigate turning circle. Additional qualitative validation was carried out using a small microcomputer based simulator, operated by an experienced frigate navigator.

The Modular Model for Simulators

It is clear that there is a growing need for simulator operators to have the ability to generate mathematical models specifically for particular tasks on the simulator. Because of the wide range of simulator tasks which need to be done, it is advisable for there to be a range of models available, which are all fully compatible with the simulator specification, and which can be used interchangeably. A further need is for simulator operators to be able to use each other's models, as it is becoming more common for a problem to be run on more than one simulator. For example, a problem in port research may be carried out using one simulator, and a training programme carried out on another. In such circumstances it is most desirable if the model can be transferred from one simulator to another.

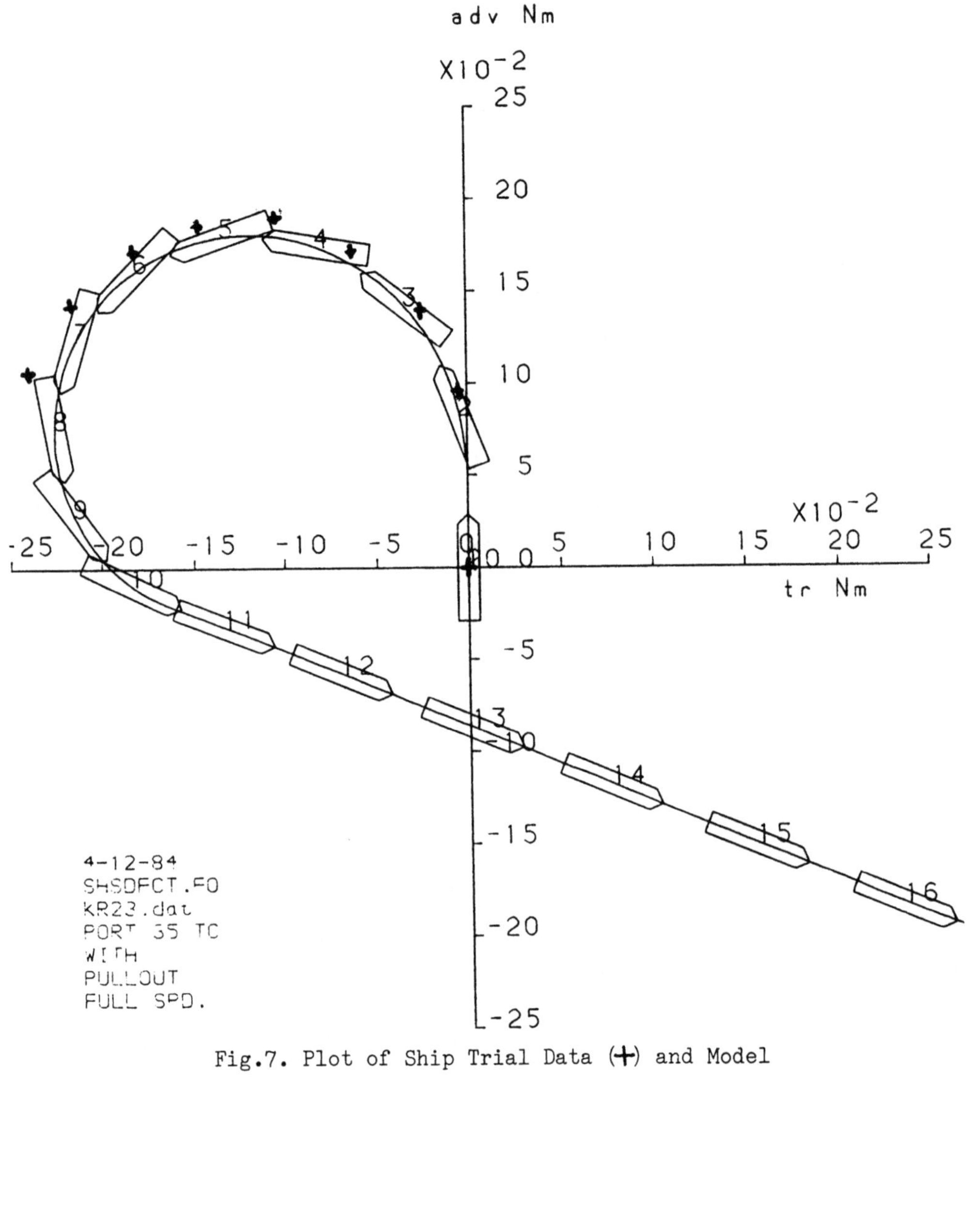

Fig.7. Plot of Ship Trial Data (+) and Model

The most important single aid to this process would be establishing a set of agreed boundaries to the mathematical model, so that a number of models could be used, all of which had the same inputs and outputs, but which could differ internally to suit the requirements of a particular user. The use of an agreed format could also greatly facilitate the transfer of a model from one simulator to another, as they would have the same internal format to accept the model. A major advantage of this model structure would be that commercial interests could be preserved, as it would not be necessary for operators to divulge the internal details of their models unless appropriate commercial arrangements were made. All that is necessary is for the inputs and outputs to be specified.

A suggestion is therefore made that the international maritime simulator community should agree on a common modular structure for mathematical models, to facilitate their common development and interchange.

References

1. System Control Inc, (USA): A system for estimation of hydrodynamic coefficients for full-scale ship tests. US Maritime Administration Pub. NMRC-207. 1981.

2. Sander, KMC.: SUSAN - a sophisticated tool for training and research in the maritime fields. Proc. 2nd International Conf. on Marine Simulation, (MARSIM 81), New York, June 1981, pp E2.1 - E2.10.

3. McCallum IR.: "Which Model?" a critical survey of ship simulator mathematical models. Proc. 3rd International Conf. on Marine Simulation, (MARSIM 84), Rotterdam, June 1984, pp 375-387.

4. Comstock PJ, (Ed.): Principles of Naval Architecture. SNAME, New York, 1967, pp 504-5

5. McCallum IR.: A ship steering mathematical model for all manoeuvring regimes. Proc. IIC Symposium on ship steering automatic control, Genoa, June 1980, pp 119 - 139.

Session 8 Off-Shore Modelling

ABAQUS/AQUA Application to Offshore Risers and Pipelines

J.F. MC NAMARA
Professor of Mechanical Engineering, University College, Galway, Ireland; currently Visiting Professor, Brown University, Providence, R. I., U.S.A;
J.P. GILROY,
Development Manager, Marine Computation Services, The Science Park, Galway, Ireland;
E.P. SORENSEN,
Hibbitt, Karlsson and Sorensen, Inc., 35 South Angell St., Providence, R. I. 02906, U.S.A; and
H.D. HIBBITT,
Hibbitt, Karlsson and Sorensen, Inc., 35 South Angell St., Providence, R. I. 02906, U.S.A

1. INTRODUCTION.

This paper outlines the capabilities of the AQUA subset of the ABAQUS (1) finite element program for simulating marine risers and pipelines under operational conditions. The AQUA suite is used mainly for nonlinear cases, the nonlinearities being due to the fluid loading and the large motions and stresses in the piping structures. Typical examples include the response due to waves and currents in deep water drilling and production risers; mooring lines; flexible risers and tanker loading hoses; tension leg tethers; offshore mining pipes; and the towing, drawdown and installation of long strings of piping. AQUA also models the J-tube pull operation, with rigid or flexible J-tubes.

The overall structure of the ABAQUS/AQUA program is described in Section 2. Section 3 discusses the background theory for the modelling features used in the illustrative example problems in Section 4. A summary is given in Section 5.

2. GENERAL STRUCTURE OF ABAQUS/AQUA.

As a subset of ABAQUS, the AQUA routines have access to all the general features and organization of the main ABAQUS data structures. The program is divided into three principal modules: a pre-processor, an analysis package and a post-processor.

The engineering analysis package is subdivided into five libraries: elements, constitutive routines, procedures, arithmetic, and data management. The libraries are organized independently, so that features such as wave loading routines are available for use with all of the elements.

Problem definition is accomplished through the pre-processor, using a keyword-parameter-data structure. The pre-processor performs extensive data checks. The user defines a problem in terms of keyword cards and "sets". The keywords introduce blocks of data, such as nodal coordinates, and may be further qualified by parameters with values given on the keyword cards. These parameters may specify logical input/output device numbers, coordinate systems, wave theory type, solution convergence tolerances, and so on. The "set" concept is very useful in organizing groups of data that have common attributes, such as material properties, boundary conditions, loading functions or collections of nodes and elements for output purposes. The "sets" may be arranged on a multi-level basis to any order.

The most critical part of problem solving with ABAQUS/AQUA is the selection of solution parameters governing the control and convergence of the nonlinear solution procedure, especially when the configuration of the structure is nearly singular (as might happen for a freely floating pipeline under a simulated towing operation). The examples described in Section 4 include pipeline problems which exhibit such difficulties.

3. DISCUSSION OF MODELLING FEATURES.

References (1) and (3) provide definition of ABAQUS capabilities. Attention here is restricted to the AQUA options.

3.1 Beam or Piping Elements.

The usual stiffness formulation is not adequate for very slender pipelines and it is necessary to use a mixed method, where axial force and, if necessary, transverse shear forces, are treated as independent unknowns (4). These "hybrid" beam elements are used for cases where the length/radius ratio exceeds 10^3. Typical problems involve very large translations and rotations, but the strains remain small.

The element library available to ABAQUS/AQUA includes beam and pipe elements based on stiffness and hybrid formulations, with linear, quadratic, and cubic interpolation. The linear and quadratic elements use independent interpolation for displacement and rotation, and include shear deformation (5). The linearly interpolated elements are particularly effective computationally.

3.2 Hydrodynamic Loads.

In addition to the usual mechanical and pressure loads available with the ABAQUS program, load terms for the offshore environment are simulated under the AQUA option. These include distributed drag, inertia and buoyancy loads, and point drag and buoyancy terms. For beam-type structures of relatively small diameter, drag and inertia forces are introduced via Morison's equation. Fluid velocities and accelerations are computed from the appropriate wave theory. The drag and inertia load vectors $\underset{\sim}{F}_D$ and $\underset{\sim}{F}_T$, respectively, are written as follows:

$$\underset{\sim}{F}_D = \frac{1}{2} \rho C_D \; D \; \Delta \underset{\sim}{v}_n \; (\Delta \underset{\sim}{v}_n \cdot \Delta \underset{\sim}{v}_n)^{1/2} \quad (1)$$

and

$$\underset{\sim}{F}_T = \frac{1}{4} \rho \pi D^2 \; [C_M \; (\underset{\sim}{a}_f - \underset{\sim}{a}_f \cdot \underset{\sim}{t} \; \underset{\sim}{t}) + (1 - C_M) \; (\underset{\sim}{a}_p - \underset{\sim}{a}_p \cdot \underset{\sim}{t} \; \underset{\sim}{t}) \;] \quad (2)$$

where

$\underset{\sim}{v}_f$ = fluid particle velocity due to steady current and waves,

$\underset{\sim}{a}_f$ = fluid particle acceleration due to waves,

$\underset{\sim}{a}_p$, $\underset{\sim}{v}_p$ = acceleration and velocity of a point on the beam,

$\Delta \underset{\sim}{v}$ = $\underset{\sim}{v}_f - \underset{\sim}{v}_p$ is the relative fluid velocity,

$\underset{\sim}{t}$ = unit vector defining the axial direction at a point in the beam

$\Delta \underset{\sim}{v}_t$ = $(\Delta \underset{\sim}{v} \cdot \underset{\sim}{t}) \; \underset{\sim}{t}$ is the relative tangential (axial) velocity of the fluid,

$\Delta \underset{\sim}{v}_n$ = $\Delta \underset{\sim}{v} - \Delta \underset{\sim}{v}_t$ is the relative transverse velocity of the fluid,

C_D = transverse drag coefficient,

C_M = transverse inertia coefficient,

and ρ = fluid density .

The total hydrodynamic load is the vector sum of Equations (1) and (2) above. The program also includes a load term for "skin friction" or tangential drag which is similar to Equation (1).

Tangential drag and inertia loads become significant at section changes (Figure 1).

The drag and inertia load vectors $\underset{\sim}{F}_{ds}$ and $\underset{\sim}{F}_{Is}$, respectively, are

$$\underset{\sim}{F}_{ds} = \frac{1}{2} \rho C_n \Delta A \Delta \underset{\sim}{v}_t (\Delta \underset{\sim}{v}_t \cdot \Delta \underset{\sim}{v}_t)^{1/2} \quad (3)$$

$$\underset{\sim}{F}_{Is} = \rho K_{ts} F_{1s} \underset{\sim}{a}_f \cdot \underset{\sim}{t} \underset{\sim}{t} - \rho (K_{ts} - 1) F_{2s} \underset{\sim}{a}_p \cdot \underset{\sim}{t} \underset{\sim}{t} . \quad (4)$$

where

s = curvilinear coordinate locating position along the structure,

ΔA = change in area of the section,

$\underset{\sim}{t}$ = unit vector pointing into the larger cross-section (Figure 1)

C_n = tangential drag coefficient of the discontinuity,

K_{ts} = tangential inertia coefficient associated with the discontinuity,

and

F_{1s}, F_{2s} = fluid and structural acceleration shape factors.

Again, the total load exerted by the fluid on the section is the sum of Equations (3) and (4).

Buoyancy loading on a riser or pipeline includes normal loads due to internal and external fluids, and the loads generated by attached floatation devices. For the former the load per unit length is

$$\underset{\sim}{F}_B = -\pi (f_1 \rho_w r_o^2 - f_2 \rho_f r_1^2) (\underset{\sim}{g} - \underset{\sim}{g} \cdot \underset{\sim}{t} \underset{\sim}{t}) +$$
$$- \pi g [f_1 (z_w - z) \rho_w r_o^2 - f_2 (z_f - z) \rho_f r_i^2] \cdot$$
$$[\underset{\sim}{n}_1 (\frac{d^2 \underset{\sim}{x}}{d\underset{\sim}{s}^2} \cdot \underset{\sim}{n}_1) + \underset{\sim}{n}_2 (\frac{d^2 \underset{\sim}{x}}{d\underset{\sim}{s}^2} \cdot \underset{\sim}{n}_2)] \quad (5)$$

where

r_o = external radius of the pipe section,

r_i = internal radius of the pipe section,

ρ_w = mass density of the external fluid,

ρ_f = mass density of the internal fluid,

z_w = free surface elevation of the external fluid,

z_f = free surface elevation of the internal fluid,

$\underset{\sim}{n}_1$, $\underset{\sim}{n}_2$ = cross-section normals as shown in Figure 1,

$\underset{\sim}{g}$ = $-g \underset{\sim}{k}$, the gravitational acceleration, $\underset{\sim}{k}$ is a unit vertical vector

$f_1 = 0$ if the elevation is above z_w and is 1 otherwise,

and

$f_2 = 0$ if the elevation is above z_f and is 1 otherwise.

At a change of section such as that shown in Figure 1, an axial buoyancy force also occurs and is given by

$$\underset{\sim}{F}_{BS} = g\,\Delta A \left[f_1\,\rho_w\,(z_w - z) - f_2\,\rho_f\,(z_f - z) \right] \underset{\sim}{t} \;, \qquad (6)$$

with z as the elevation of the section location. Buoyancy material attached externally to the pipe is assumed to exert a vertical load equal to the volume of fluid displaced.

The program offers linear Airy and 5th order Stokes wave theories (7), from which fluid particle velocities and accelerations are computed for the hydrodynamic loading routines.

3.3 Friction, Surface Contact and Interface Elements.

Friction forces arise in problems where there is contact between a pipeline and the seabed. The frictional resistance usually is higher for motion transverse to the pipe then for motion along the pipe. Coulomb friction with an elliptic flow surface is used to simulate this (4). The contact pressure between the surfaces is defined as an exponential function of their interpenetration.

Several interface formulations are provided, from the simplest case with infinitesimal relative shear motions to finite sliding between a deforming structure and a rigid surface, or between two deforming structures.

4. EXAMPLES.

The examples below illustrate the application of ABAQUS/AQUA to typical offshore problems. Additional examples are contained in (1) and (4).

4.1 Disconnected Riser.

Figure 4 is a schematic diagram of a disconnected riser. Under extreme weather conditions the riser must be unlatched and hung freely. The numerical data are listed in Table 1 and are taken from (8, 9, 10). The major modelling requirements include a flex joint 12.2 m below the vessel

TABLE 1 DATA FOR DISCONNECTED RISER

Young's Modulus, E	2.1×10^{11}	N/m^2
Mass density of steel	7850.0	kg/m^3
Mass density of internal fluid	1025.0	kg/m^3
Mass density of buoyancy material (foam)	540.0	kg/m^3
Mass density of seawater	1025.0	kg/m^3
Internal riser diameter	441.0	mm
External riser diameter	473.0	mm
Drag diameter	1054.1	mm
Buoyancy diameter (riser tube)	473.0	mm
External diameter of buoyancy module	1054.1	mm
Drag coefficient, C_D	1.0	
Inertia coefficient, C_m	2.0	
Length of riser	1219.2	m
Mass of Lower Marine Package (LMRP)	75,988	kg
Torsional spring constant for flex joint	1.94206×10^6	Nm/rad
Clearance between riser and seabed	12.2	m
Water depth	1219.2	m
Flex joint position from seabed	1219.2	m
Additional buoyancy of LMRP	97,330	N
Number of elements	25	
Time increment	0.125	secs
Current velocity, linearly decreasing from MWL to zero at 910 m from seabed	0.524	m/sec
Wave height	18.6	m
Wave period	14	secs
Vessel surge amplitude	5.486	m
Surge phase angle	73°	
Surge period	14	secs
Vessel heave amplitude	3.658	m
Heave phase angle	0°	

suspension point, discrete foam buoyancy modules, lower marine package mass, buoyancy forces on the free end, and storm wave and vessel motion input.

The major concerns are the possibility of compressive forces developing in the riser due to different responses of the vessel and riser in waves; excessive lateral rotations of the riser leading to the possibility of moonpool impact, and steep gradients of the stresses in the riser near the vessel suspension point. Results for the axial force at the support point are given in Figure 3, and for the moment 32.9 m below the flex joint are shown in Figure 4. Good agreement is obtained with reference (10). An earlier study (8) assumed rigid body vertical motion following the wave cycle, using the uncoupled translational equations of motion only. That resulted in a maximum riser kink across the flex joint of 33^0, whereas ABAQUS and (10) both indicate a maximum kink of 8^0, with 12^0 absolute rotation of the portion of the riser between the vessel and the flex joint.

4.2 Flexible Riser Analysis.

Flexible risers are constructed of complex lay-ups of steel, plastic and fabric, resulting in a non-linear moment-curvature relationship for the section.

Figure 5 shows a flexible riser connecting a subsea tower that is 150 m below the MWL (and away from the wave zone) to an oil tanker on the free surface. The riser is assumed to be full of water. The static analysis is carried out in two stages. First, the riser is allowed to sag under its own weight in water as it is held horizontally (as shown in Figure 5). This is the most difficult part of the analysis numerically. The tanker end is then moved to the MWL. The end reactions and support angles are listed on Figure 5. Finally, a dynamic analysis is executed about the at-rest position, with a single wave harmonic of height 13.6 m and 14 second period. The vessel motion is assumed to be a 2.0 m surge amplitude at 14 seconds period. The bending moment at the point of maximum curvature is given in Figure 6.

4.3 Pipeline Drawdown.

In this example, shown in Figure 7, a pipeline/cable assembly is drawn down to a point on the seabed. The pipe and cable are negatively buoyant with attached, discrete, buoyancy devices.

"Hybrid" beam elements are used for the cable and the pipeline. The extreme flexibility of the surface configuration causes severe numerical difficulties, which are avoided by moving the pipeline and cable to a less singular configuration in a rigid body motion without external loads (the "initial configuration" in Figure 7). The free end of the pipe is then drawn down, pipe and cable weights are applied, discrete buoyancy devices are attached and the cable/pipeline interface at node 3 is moved to a new position at 195 m depth, as shown in Figure 7. Node 3 is then released and node 1 moved to its final configuration, which is shown in Figure 7.

4.4 J-Tube Pull.

J-tube pulls are sometimes used to raise a pipeline to the deck of a jacket structure. The modelling features of this problem include the deformations of the J-tube and pipeline, the sliding of the pipeline through the J-tube, and the simulation of the winching cable. This example uses "slide lines" to model the interaction of the pipeline and J-tube.

The numerical data and the initial configuration of the assembly are shown in Figure 8. The problem is simplified by considering the pipeline to lie initially within a horizontal extension to the J-tube. Both the pipe and the J-tube are modelled with pipe elements that include hoop stress effects due to uniform expansion and contraction of the pipe; the cable is modelled with first order hybrid beam elements. The pipeline/J-tube interface includes Coulomb friction and an exponential normal contact pressure versus clearance distance relationship.

The anaylsis is performed in two steps. Firstly, deadweight is applied to the pipeline, the horizontal extent of the J-tube is lowered by 0.254 m, and a back tension of 447.3 N is applied to the end of the pipeline to represent the tension of the pipeline on the seafloor. Then, the pipeline is winched through the J-tube by specifying the position of the cable end. The cable force is shown as a function of total pull distance in Figure 9. The final configuration of the pipeline is shown in Figure 10.

5. Closure.

The capabilities of the AQUA subset of ABAQUS have been discussed, and some simulations have been presented. The breadth of capability indicated by these examples suggests the value of a well designed, general purpose code where the basic algorithms have been extensively developed on a wide range of practical applications.

6. REFERENCES.

(1) ABAQUS manuals, Vol. 1; User's manual, Vol. 2: Theory manual, Vol 3: Example Problems manual, Vol. 4: Systems manual, published by Hibbitt, Karlsson and Sorensen, Inc., Providence, R. I. (1984).

(2) Skjelbreia, L. and Hendrickson, J., "Fifth Order Gravity Wave Theory", chapter 10 of the proceedings of the 87th Coastal Engineering Conference, den Haag, (1960).

(3) Hibbitt, H.D., "ABAQUS/EPGEN - A General Purpose Finite Element Code With Emphasis On Nonlinear Applications", Nuclear Engineering and Design, 77 (1984) pp. 271-279.

(4) Hibbitt, H.D., Becker, E.P., and Taylor, L.M., "Nonlinear Analysis of Some Slender Pipelines", Computer Methods in Applied Mechanics and Engineering, 17/18 (1979) pp. 203-225.

(5) Hughes, T.J.R., Taylor, R. L. and Kanoknukulchai, W., "A Simple and Efficient Element for Plate Bending", Computer Methods in Applied Mechanics and Engineering, 11/10, (1977) pp. 1529-1543.

(6) McNamara, J.F. and Lane, M., "Practical Modelling for Articulated Risers and Loading Columns", Journal of Energy Resources Technology, Trans. ASME, 106, (1984) pp. 444-450.

(7) Sarpkaya, T. and Isaacson, M., "Mechanics of Wave Forces on Offshore Structures", Van Nostrand Reinhold Company, New York, (1981) ISBN 0-442-25402-4.

(8) Kwan, C.T., Marion, T.L. and Gardner, T.N., "Storm Disconnect of Deepwater Drilling Risers", Proceedings of the 11th Annual Offshore Technology Conference, Paper OTC-3586, Houston, Texas, (May 1979) pp. 2001-2009.

(9) McNamara, J.F., Joyce, P.J. and Gilroy, J.P., "Arbitrarily Large Static and Dynamic Motions of Drilling and Production Risers", Proceedings of the International Symposium on Hydrodynamics in Ocean Engineering, Norwegian Hydrodynamic Laboratories, Tondheim, Norway, (August 1981) pp. 865-886.

(10) McNamara, J.F. and Gilroy, J.P., "Nonlinear Dynamics of Disconnected Marine Risers", Proceedings of the 4th International Symposium on Offshore Mechanics and Artic Engineering, Dallas, Texas, (February 1985).

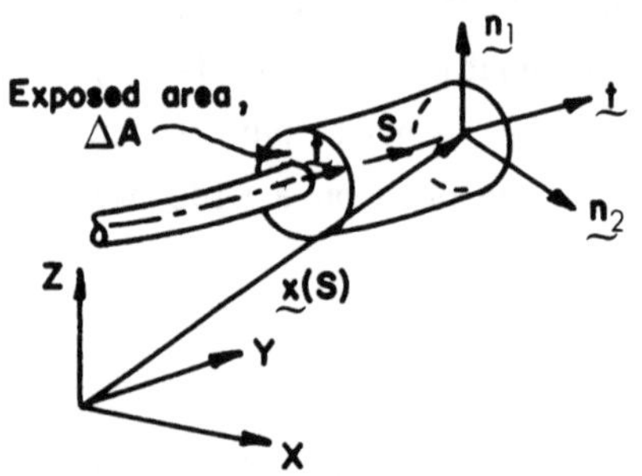

Fig. 1 Notation for hydrodynamic loading.

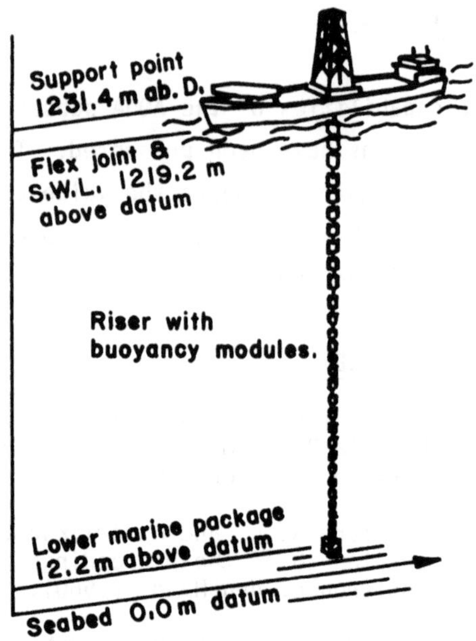

Fig. 2 Storm disconnect riser with lower marine package.

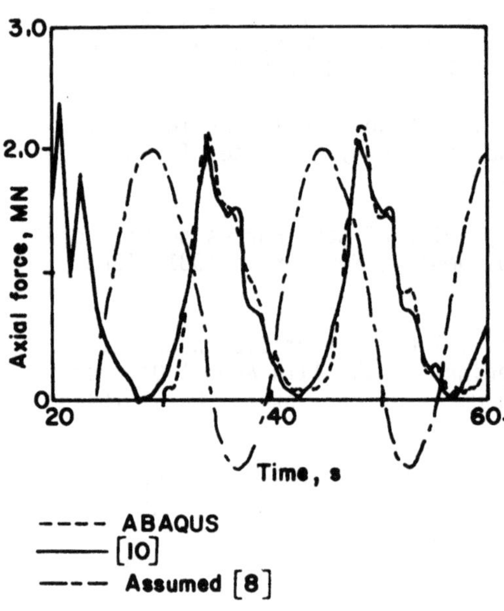

Fig. 3 Axial force in hanging riser at vessel support point.

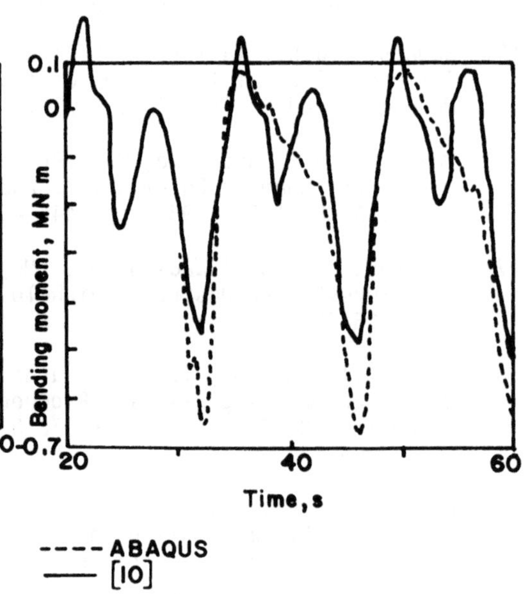

Fig. 4 Bending moment in hanging riser at 39.2 m below flexible joint.

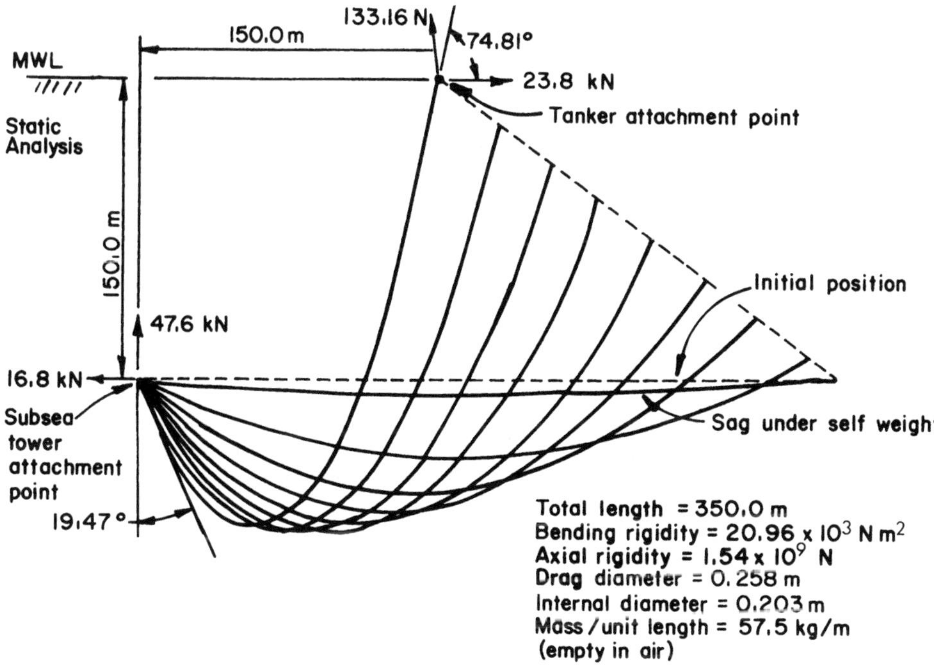

Fig. 5 Transfer of flexible riser to surface.

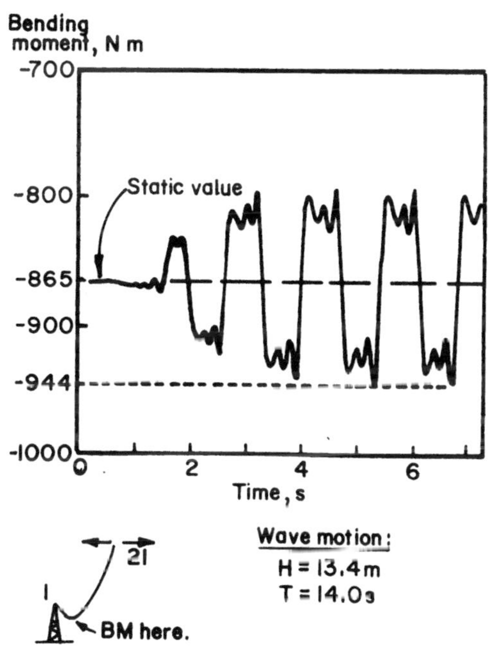

Fig. 6 Bending moment at point of maximum curvature.

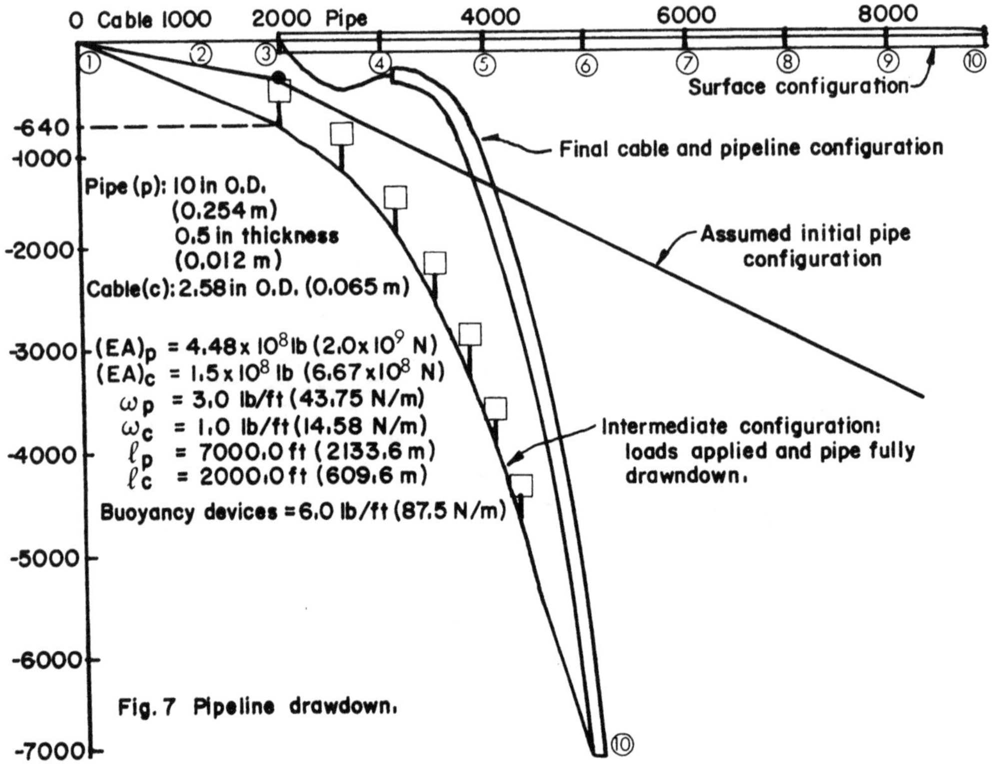

Fig. 7 Pipeline drawdown.

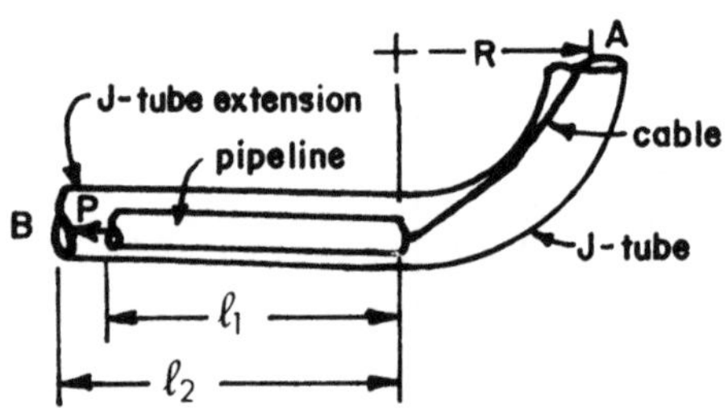

Geometry:

R = 12.68 m (41.6 ft)

ℓ_1 = 30.48 m (100.0 ft)

ℓ_2 = 38.10 m (125.0 ft)

J-tube outer diameter = 0.4572 m (1.5 ft)

J-tube thickness = 0.0191 m (0.0625 ft)

Pipeline outer diameter = 0.3239 m (1.062 ft)

Pipeline thickness = 0.0159 m (0.0521 ft)

Concentrated force:

P = 4447.3 N (1000.0 lb)

Friction coefficient between cable/pipeline and J-tube = 0.3

Fig. 8 Problem specification: simplified J-tube analysis.

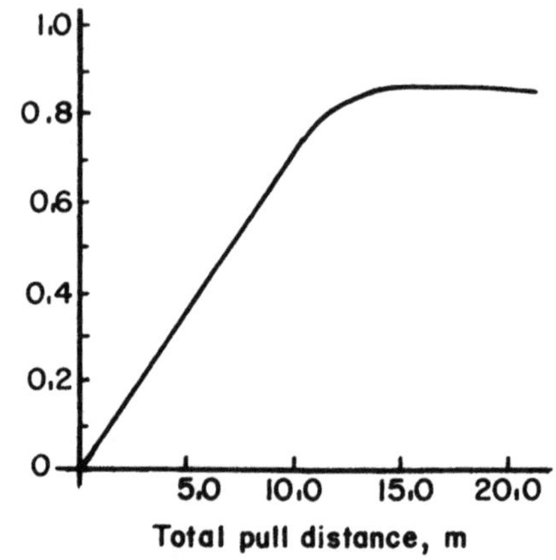

Fig. 9 Cable force versus total pull distance for J-tube pull.

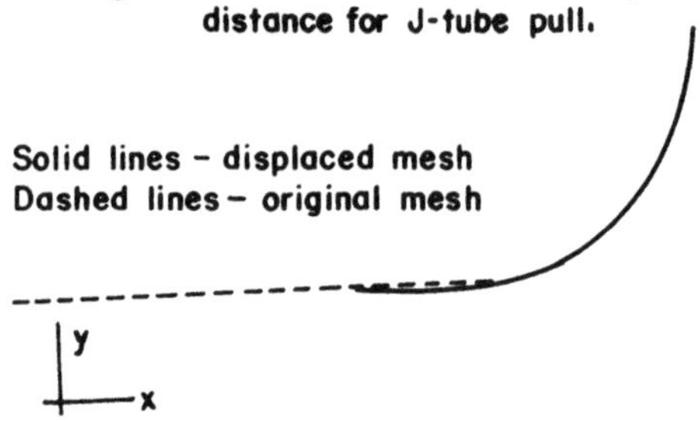

Fig. 10 Pipeline configuration after pull up through flexible J-tube.

Session 9 Steering and Control of Marine Vehicles

Steering Control Algorithms for a Submersible

L.A.M.Kuijk, G.B.H.Jacobs and B.Twigt

Applied Dynamics Europe,
a department of
Van Rietschoten & Houwens E.M. B.V.
Rotterdam, The Netherlands

ABSTRACT

This paper describes the algorithm design of a steering system for control of the depth and the course of a manned submersible.

The control algorithms will be implemented into a general purpose military computer system which enables application of control techniques such as :

- optimal linear quadratic control;

- feed forward control and mode switching;

- gain scheduling for the speed;

- adaptive Kalman filtering;

- estimation of slowly varying disturbances.

The decoupled course and depth control algorithms use simplified control models, respectively based upon a fourth order linear approach of the depth behaviour and an extended first order Nomoto model for the course behaviour.

1. INTRODUCTION

Although in the design of a steering control system the control algorithms play a minor part, compared with hardware design and implementation, this paper is limited to the control strategy.
The work that V. Amerongen (Ref. 1) has done in the past, and is still doing in this field has been of great importance to us. The use of modern simulation equipment proved to be very advantageous.
The submersible for which the algorithms are developed can be controlled by means of two sets of hydroplanes for depth and pitch control, and one rudder for course control. A good approximation of the hydrodynamic parameters of the vessel are known, both via model tests and trials with a similar type of submersible.

2. MODELLING THE SHIP

In order to develop and test the control algorithms in a realistic environment , the first goal was to simulate the complete 6 DOF equations of motions of the submersible as accurate as possible.
The equations of motion as given in Report 2510 of the Naval Ship Research and Development Centre were used. The hydrodynamic coefficients of the non-linear model were computed in order to get a first approximation of the model. Since the design of the submersible and the computation of the hydrodynamic coefficients was the responsibility of the yard, our work in this modelling work was limited to the validation process.
Full scale trials were done in the past with a similar type of submersible. Measurements made during those trials were available. The results of the simulation were compared with the measured responses of the submersible during the sea trials. Although in general a good agreement between the simulated and the actual manoeuvres was found some refinements in the model parameters appeared to be necessary.
The hydrodynamic coefficients which are the most important for the accuracy of the model were varied in the simulation and the effect on the model was determined. During the sea trials of the submersible with the v.R&H steering control system again a number of the standard manoeuvres will be executed. The then measured data will again be compared with the simulation data. This, together with the information on the sensitivity of certain coefficients, should enable us to refine the model when large discrepancies exist between model and real data. The mathematical model, thus obtained, is more than adequate in order to be used as a test vehicle for various control algorithms.
The model becomes inadequate only if the angle of attack of a control surface becomes large, causing a condition of stall in which the lift no longer increases with the deflection angle. In normal submerged manoeuvres, including steep dives and tight turns stall will not be experienced.
From this model,simplified models for depth and course were derived which are independent of each other under the following assumptions:

- An appoximately symmetrical submersible;

- Small roll angles.

For the vertical plane, the equations of this model are:

$$\begin{bmatrix} \dot{z} \\ \ddot{z} \\ \dot{\theta} \\ \ddot{\theta} \end{bmatrix} = \begin{bmatrix} A \end{bmatrix} \begin{bmatrix} z \\ \dot{z} \\ \theta \\ \dot{\theta} \end{bmatrix} + \begin{bmatrix} B \end{bmatrix} \begin{bmatrix} db \\ ds \end{bmatrix} + \begin{bmatrix} C \end{bmatrix} \begin{bmatrix} F \\ M \end{bmatrix}$$

The terms of the A,B and C matrices are a function of the hydrodynamic coefficients and (some) of the forward speed of the submersible.
The outputs of this simplified model, depth and pitch, are close to those of the complete 6 DOF model, even when large hydroplane angles are commanded. This model is used in several places in the controller.
It is used in order to get noise free estimates of the states of the submersible, and it is used in order to compute the low frequent disturbances which act upon the submersible, such as mis-ballast and mis-trim situations.

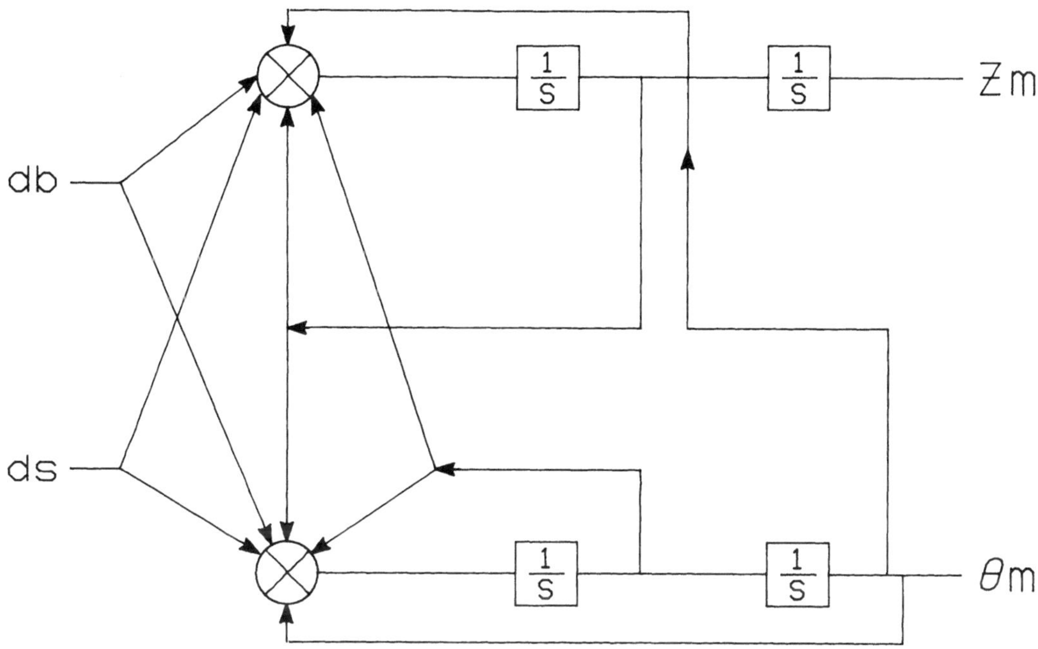

Fig. 2 Simplified model for the vertical plane

For the horizontal plane, a model is derived which is valid for most rudder angles and the whole (positive) speed range.
This model can be used for both course keeping and for course changing manoeuvers. From the two equations which describe the motions of the 6 DOF model in the horizontal plane can be computed that the lateral speed, v, has an approx. linear relation to the course rate of change. The equation for the lateral displacement, under the assumptions that the angular velocity components relative to fluid about X and Y axis, p and q, are zero and that all accelerations are zero, simplifies to:

$$h_1 * u * r + h_2 * r * |v| + h_3 * u * v + h_4 * dv * u * u + h_5 * v * |v| = 0$$

substitution of $v = C * \dot{\psi}$, $r = \dot{\psi}$ and the values for the hydrodynamic coëfficients results in:

$$-dv = \frac{1}{u} * \dot{\psi} + 60 * \frac{1}{u} * \frac{1}{u} * \dot{\psi} * |\dot{\psi}|$$

substitution of $r' = \frac{1}{u} * L * \dot{\psi}$ results in:

$$-dv = \frac{1}{L} * r' + \frac{60}{L * L} * r' * |r'|$$

This expression gives a good description of the relation between the rudder dv and the variable r during course changing operations. The equation for the yawing moment, under the assumption that $v=\dot{v}=0$, is:

$$Iz * \dot{r} = h_6 * \dot{r} + h_7 * r + h_8 * dv$$

This is the well known first order Nomoto model, which is only valid, because of the assumption that was made, for small rudder angles. Combining this Nomoto model with the simplified equation for the lateral displacement results in a model which is valid for all rudder angles and the whole speed range.
The disadvantage of this model is that non linearities exist, which prohibit the use of optimal linear feed back control. However, during course changing operations the rudder angle will be largely determined by the feed forward control, and during course keeping a suitable Kp and Kd for a PD type of controller are given by V. Amerongen (Ref. 1)

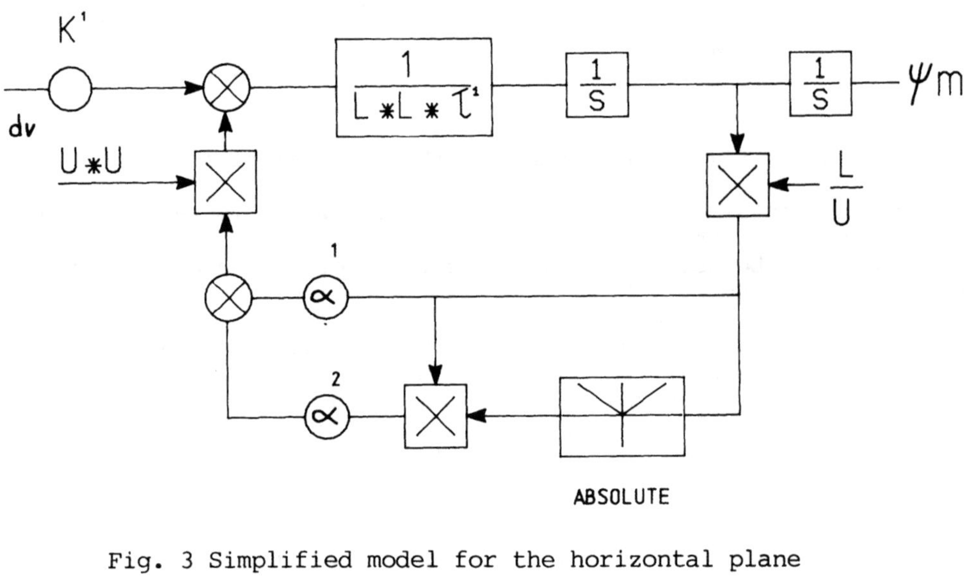

Fig. 3 Simplified model for the horizontal plane

3. MODELLING THE DISTURBANCES

The forces acting upon the submersible, due to waves, consist of the first order wave forces and the second order (low frequency) forces. Especially at periscope depth these forces are important and they have to be included in the simulation in order to check the control system.
It is usual to generate the first order motions of the submersible instead of the first order forces. The forces and moments in regular waves can be computed by means of programs based upon strip theory. With these results available, the forces and moments in irregular waves can be computed as well. Normally this is done off-line by setting up files containing time histories of the submersible's high frequency (first order) motions and low frequency (second order) drift forces and moments. Files are set up for various wave, heading and speed conditions. If freely varying speed, sea and heading corrections are required, the motions and forces are best calculated on-line, using a reduced number of wave components. These calculations are very time consuming and require large data tables to be used for function generation. Since the simulation equipment we use is capable of performing the function generation task all motions, forces and moments are computed on-line. An energy density function, which describes a particular sea state, often used in open water is the Pierson Moskowitz spectrum defined as:

$$S(\omega) = \frac{A}{\omega^5} e^{-\frac{B}{\omega^4}}$$

$$A = 0.77915, \quad B = 3.11/Hs*Hs$$

Hs is the significant wave height.
This spectrum is divided into 24 bands, each with the same frequency interval dw, each band can be associated with an individual wave having the same energy and frequency as the band. The wave elevation can be described by a combination of the 24 waves using random phase angles e(i). The amplitude of a wave with frequency w(i) is:

$$Ai = \sqrt{2S(\omega i)d\omega}$$

The wave elevation at the centre of gravity of the vessel is:

$$Z(t) = \sum_{i=1}^{24} Ai\cos(\omega i\,t(1-\omega i\,v/g)+ei)$$

where wi(1-wi*V/g) is the encounter frequency of the waves. This value Z(t) is added as noise to the depth from the model. Each wave component generates a motion in heave, pitch and yaw. The amplitude and phase of the motion are given in tables. By means of function generation with the wave direction and the speed of the submersible as inputs, the first order displacements are computed. The squared instantaneous wave amplitude, A*A, and the instantaneous wave frequency, W, are also computed. The drift forces and moments are given by Fi(W)*A*A, were Fi is a table with either heave forces, pitch moments or yaw moments. Since all forces and moments are given in an earth fixed coordinate system, they have to be transformed to the body axis system of the submersible and then used in the 6 DOF equations of motions of the submersible.

4. CONTROL STRATEGY IN THE VERTICAL PLANE

The control of the submersible takes place on several levels. The first level consists of a linear feed back of the states, based upon optimal control theory. This feed back control reduces the state errors by ordering the proper plane deflections as a function between the reference state and the actual state. The actual state is the output of the linearized model, which first of all filters the measured values of pitch and depth and gives an estimation of the pitch rate of change and the depth rate of change. The feed back gains are computed in such a way that a quadratic cost function is minimized. This cost function is chosen according to the specifications of the autopilot and the obtained simulation results. The specifications of the autopilot ask for minimal state errors and for minimal hydroplane activities.
Since these specifications are contradictory, two sets of feed back gains are computed by means of two cost functions. One cost function is used for feed back gains where the state errors are minimised, with the penalty that larger plane angles are allowed. The other cost function minimizes the plane activities with as penalty that larger state errors will occur. The user of the system has the choice which of the two sets of gains is used, only during depth changing manoeuvres the system itself will switch to the set where the state errors are minimized.
Before the ordered plane deflections from this optimal controller are actually used, they are checked against the possibilities of the steering machines. If the ordered plane speeds exceed the speed limits of the steering machines, the plane orders are adjusted in such a way that the commands are in agreement with the speed of the steering machines. In order to compute the eight feed back gains (four for each plane since the state error vector consists of depth and pitch error and depth speed and pitch speed error) it is necessary to determine a linear model in the vertical plane of the submersible. Some of the terms in the A, B and C matrices of the model mentioned in the previous chapter depend upon the speed of the submersible. In order to get matrices with constant coefficients needed to compute the gains, the speed range of the submersible is divided into a number of intervals and for each speed interval the eight gains are computed by linear interpolation.

The second level of control consists of a feed forward controller. For depth changes it is possible to compute the necessary plane angles in order to achieve the desired values of pitch angle and depth rate of change.
The assumption that w=0 and that $p=q=r=\dot{p}=\dot{q}=\dot{r}=\dot{u}=\dot{v}=\dot{w}=v=0$ simplifies

the non-linear equations for the normal force and the pitching moment into:

$$z_1 * ds + z_2 * db = 0$$

$$m_1 * \theta + m_2 * ds + m_3 * db = 0$$

From these two equations the values for the plane angles during major depth changes were w=0, db and ds, can be computed. These plane angles are sent directly to the steering machines. This feed forward control is also used in cases that a certain pitch angle is desired during depth keeping. Again the assumption is that $p=q=r=\dot{p}=\dot{q}=\dot{r}=\dot{u}=\dot{v}=\dot{w}=v=0$, but now $w \neq 0$. The same before mentioned equations now become:

$$z_1 * ds * u * u + z_2 * db * u * u + z_3 * u * w + z_4 * w * |w| = 0$$

$$m_1 * \theta * u * u + m_2 * ds * u * u + m_3 * db * u * u + m_4 * u * w = 0$$

Substituting $w = u * \theta$ and dividing by $u * u$ results in :

$$z_1 * ds + z_2 * db + z_3 * \theta + z_4 * \theta * |\theta| = 0$$

$$m_1 * \theta + m_2 * ds + m_3 * db + m_4 * \theta = 0$$

From these two equations the values for the hydroplane angles can be computed in order to maintain this desired pitch during depth keeping.

The third level of control consists of the computation of the mean values of disturbing forces and moments and the necessary plane angles to counteract those values.
For this purpose a second internal model is used, which has as input the measured plane angles of the submersible. Disturbing force and moments are not present in this model. The difference between the outputs (pitch and depth) of this model and the measured pitch and depth is used to update the model in order to prevent this model from drifting to far from the states of the real submersible. By combining the model values with this difference, an estimation is achieved of the depth- and pitch rate of change of the real submersible. The linear equations for the vertical plane are given by :

$$\ddot{z} = a_{22} * \dot{z} + a_{23} * \theta + a_{24} * \dot{\theta} + b_{21} * ds + b_{22} * db + c_{21} * F$$

$$\ddot{\theta} = a_{42} * \dot{z} + a_{43} * \theta + a_{44} * \dot{\theta} + b_{41} * ds + b_{42} * db + c_{41} * F + c_{42} * M$$

The equations for the model do not contain F and M, the plane angles for the model are the measured angles minus the correction angles.

$$\ddot{z}_m = a_{22} * \dot{z}_m + a_{23} * \theta_m + a_{24} * \dot{\theta}_m + b_{21} * (ds - \Delta ds) + b_{22} * (db - \Delta db)$$

$$\ddot{\theta}_m = a_{42} * \dot{z}_m + a_{43} * \theta_m + a_{44} * \dot{\theta}_m + b_{41} * (ds - \Delta ds) + b_{42} * (db - \Delta db)$$

substitute $\Delta ds = \Delta sds + \Delta\Delta ds$, $\Delta db = \Delta sdb + \Delta\Delta db$

Subtracting the model equations from those of the submersible gives:

$$a_{22} * \Delta \dot{z} + a_{23} * \Delta \theta + a_{24} * \Delta \dot{\theta} + b_{21} * \Delta\Delta ds + b_{22} * \Delta\Delta db = 0$$

$$a_{42} * \Delta \dot{z} + a_{43} * \Delta \theta + a_{44} * \Delta \dot{\theta} + b_{41} * \Delta\Delta ds + b_{42} * \Delta\Delta db = 0$$

These equations can be solved for $\Delta\Delta db$ and $\Delta\Delta ds$. Integrating these values gives the offset plane angles Δ db and Δ ds. Due to waves, the estimated depth rate of change and the pitch rate of change vary with the waves. Therefore those values are first low-pass filtered before they are used in the steering machines.

5. CONTROL IN THE HORIZONTAL PLANE

The first level of control consists of a PD type of controller. For the proportional part of the controller, Kp is chosen as :

$$K_p = -2.5 * \frac{u_0}{u}, \quad -K_p \leq 5$$

were u0 is the cruising speed of the submersible. According to V.Amerongen (Ref. 1) a suitable Kd is :

$$K_d = \frac{L}{u} * (2 * d * \sqrt{K_p K' \tau' - 1}) / K'$$

were K' and τ' are gain and time constant of the Nomoto model. The damping, d, is determined by means of simulation.

A second level of control is a feed forward controller. The relationship between rudder angle dv and the desired course rate of change is derived from the equation for the lateral displacement of the submersible.

$$-dv = \frac{1}{L} * r' + \frac{60}{L * L} * r' * |r'|$$

The third level of control compares the observed course rate with that of an internal model. The difference between those values is used to compute the yawing moment and the offset rudder.

6. THE CONTROLLER MODES

Both depth and course controllers distinguish between two modes, one for course (depth) changing, and one for course (depth) keeping. During the keeping mode the rudder/plane angles are mainly generated by the optimal feed back controller, during the changing mode it is mainly the feed forward controller which generates the angles. It is clear that it is desirable that no large rudder/plane jumps should occur when the control switches its mode.
For the course controller the rudder angle in the changing mode is:

$$dv = \alpha_1 * \dot{\psi}g + \alpha_2 * \dot{\psi}g * |\dot{\psi}g|$$

When the mode is switched to keeping, the PD controller generates in the first instant:

$$dv' - K_p * \Delta\psi + K_d * \dot{\psi}g$$

were $\dot{\psi}g$ = desired rate of change
$\Delta\psi$ = heading error. The mode switching is optimal when $dv = dv'$ or :

$$\Delta\psi = (\alpha_1 - K_d + \alpha_2 * |\dot{\psi}g|) * \dot{\psi}g / K_p$$

since $\alpha_1 << K_d$ and $\alpha_2 * |\dot{\psi}g| << K_d$, this results in:

$$\Delta\psi = -\frac{K_d}{K_p} * \dot{\psi}g$$

When during course changes the heading error becomes $\Delta\psi$, the mode is switched to keeping.
For the depth controller, the optimal switching point is somewhat different for sailplane and sternplane. Averaging those two values gives satisfactory results.

7. FILTERING AND ESTIMATION OF STATE VARIABLES

Besides low frequency disturbances which act upon the submersible, such as mis-trim and mis-ballast moments and forces and the second order wave forces and moments, which can be counteracted, there are also higher frequency disturbances. Examples are the influence of the waves on the measured depth and the first order wave forces and moments. The resulting effect is that high frequency rudder and plane orders are generated by the controller, which have no positive effect on the reduction of the setpoint errors. Therefore it is necessary to filter the measured signals in order to suppress these high frequency motions.
For this purpose the before mentioned model of the submersible is used in the control computer. This model is actuated by the actual measured plane (rudder) angles and the calculated offset plane angles. The states of this model and the measured states of the submersible are weighed - in a certain proportion - in order to get filtered system states. The low frequency energy, and the total energy of the difference between measured state and model state is computed, and dividing those two values gives the weighing factor. This weighing factor multiplied by the error signal, is also used as feed back signal in the model. This model is now a noise adaptive Kalman type filter. Besides filtering the measured states, it also provides estimates of the derivatives of those states. In fig. 4 is illustrated how the computed feed back gain is used in order to get filtered system states. The feed back value is the noise ratio (KSID) multiplied by the difference between measured depth, Z and model depth Zm. A filtered estimate of the depth is:

$$\hat{Z} = KSID * (Z - Zm) + Zm$$

For low values of KSID, or much high frequency noise, the estimate mainly depends upon the model depth Zm. When there is no noise, KSID =1, the estimate will equal the measured depth. In order to get the depth rate of change, the same feed back value is added to $\dot{Z}m$, the depth rate of change of the model.

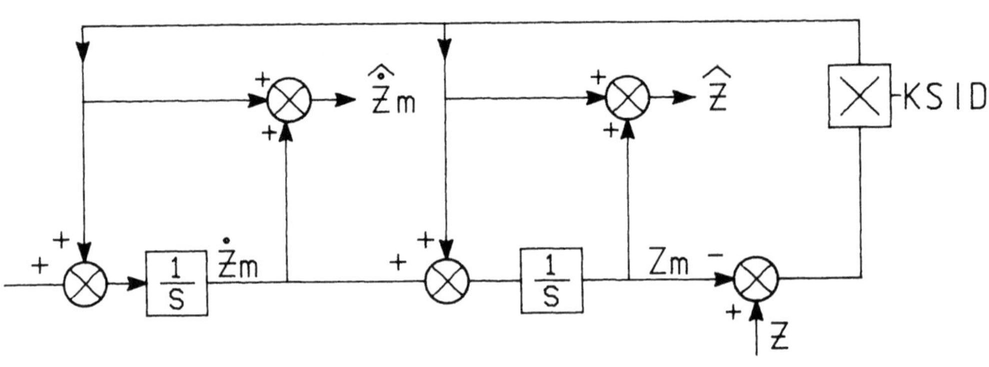

Fig. 4 Use of the weighing factor in order to get filtered states.

8. DETERMINATION OF THE NOISE RATIO IN THE FILTER MODEL

The low frequency energy component from the difference between the measured state variable and the corresponding model state variable is determined by filtering this error signal in a low pass filter with a variable, helmsman definable, time constant. The absolute value of this result is computed, then avaraged and finally squared. The total energy is computed by averaging the absolute value, and squaring the result. Dividing the low energy component by the total energy component results in the feed back gain and the weighing factor of the filter model.

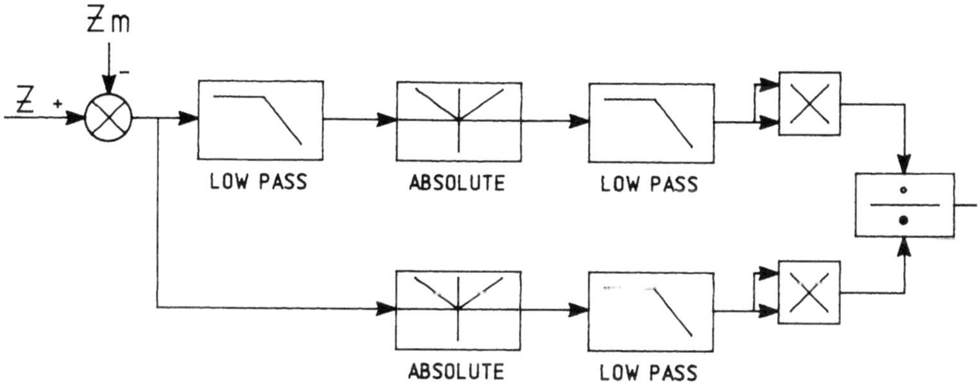

Fig. 5 Computation of feed back gain for the depth filter model.

9. SIMULATION

In order to test the control algorithms, use has been made of various computer systems. As said before the first goal was to simulate the complete 6 DOF equations of motions of the submersible. The next step was to incorporate the controllers in the simulation and the effect of waves and other disturbances. For detailed engineering purposes we used a PDP-11 system together with PSI, an interactive simulation package developed by Delft University of Technology. This system has the capacity to simulate the 6 DOF model and various parts of the controllers slightly faster than real time. Due to the limited amount of memory which can be used by a single task in the PDP-11 it is not possible to simulate all the disturbances as well. This problem was partially solved by running the PSI software on a Tulip micro processor. This microprocessor is even faster than a PDP-11 which normally has to accommodate more users.
A third system which is used consists of a PDP-11 together with an AD-10 special purpose simulation computer. With this system it is possible to simulate the 6 DOF model, all the controllers and the disturbances about 200 times faster than real time. The simulation includes, but is not limited to, 164 functions of one variable, 165 functions of two variables, 280 summers,4 coordinate transformations,12 divisions, 120 integrators and 37 combined comparator/switches. The frame time needed for one solution of all algebraic variables and the state variables is about 2 milliseconds.

The use of different systems proved to be highly advantageous. Apart from the high speed of the AD-10, comparing results of the systems immediately pointed to the small programming errors which otherwise might have stayed unnoticed. At the moment all PDP-11/AD-10 software is transfered to a new simulation system consisting of a VAX-750 AD-100 combination. With this system we hope to speed up the simulation to approx. 1000 times real time. This gives the opportunity not only to simulate various manoeuvres but complete missions as well.
A next step will be to simulate positioning equipment and to feed this information to the autopilot.
This will give the submersible the capability to sail automatically from A to B without needing manual input.

10 REFERENCES

(1) J. van Amerongen, "Adaptive Steering of Ships", Delft University of Technology, 1982

(2) Naval Ship Research and Development Center, "Report 2510"

(3) Applied Dynamics International, "six-Degree-of-Freedom Flight Simulation Using the System 10", Application Report

Session 10 Training and Traffic Control

Simulation and Pilot Training at the Panama Canal

CAPTAIN ROBERT D. VALENTINE MNI

President, International Maritime Consultants, Inc, Palm Harbor, Florida, USA
Former Chief, Pilot Training Unit, Panama Canal Commission

SUMMARY

Several years ago, the Panama Canal Commission began to have concern about its ability to adequately train pilots in the future. There were several sound reasons for this disquietude. The emergence of the shiphandling simulator appeared to offer relief. Following a somewhat exhaustive investigation, the Canal management elected to tentatively supplement the regular training with a course of instruction on a shiphandling simulator. The reasons for the choice of facilities as well as the problems in the development of a viable and acceptable (to the pilots) program are discussed. The present status of the program and the role of the micro-simulator in that program are also presented in this paper.

BACKGROUND

As ships began to increase in size, the spectre of pollution and other disasters began to create considerable pressure on the maritime industry. In the parliaments and congresses of the maritime nations the concern for improved maritime safety focused, in part, on training. International bodies such as the I.M.O., I.C.S. and OCIMF echoed these concerns. During this period tremendous technological advances occurred. The application of this technology led to the development of sophisticated simulators which promised to be the premier training tool of the future. The authorities at the Panama Canal began to take note of these developments.

By the 1980's the shiphandling simulator had gained acceptance by the maritime industry for training bridge personnel in teamwork, voyage planning and for the transition of masters and mates of conventional size vessels to larger vessels.

However, the use of the simulator in the training of pilots was not generally accepted by the pilot community. Historically, pilots had been successfully trained on the job. Nevertheless, the Panama Canal Commission (PCC) began to inquire into the use of this form of training. Consequently, the author's report to the Canal authorities following visits to several training and manufacturing facilities in the United Kingdom sustained that interest.

The Panama Canal's interest was not simply an attempt to keep in step with the advances of technology, but one born out of necessity. The changing traffic pattern at the Canal with its percentage increase of larger vessels coupled with a greater diversity in vessel types made it increasingly difficult to assign the Pilots-in-Training and the Limited Pilots to vessels of a size appropriate for training or commensurate with their experience. This difficulty generated concern on the part of the Canal authorities about their ability to insure that new pilots would be adequately prepared to move up to vessels of greater tonnage. The terms of the Panama Canal Treaty which requires the Panama Canal to hire and train Panamanian nationals at deliberate speed has caused, in the case of pilots, additional concern due to the extremely small pool of experienced Panamanian mariners from which to recruit.

INVESTIGATION AND SELECTION OF FACILITY

A second report by the author after visiting the principal simulator training facilities in the U.S. suggested a need for an in-depth study into this subject. A seven man team was formed to evaluate these sites. The study group comprised the Deputy Marine Director, three retired pilots each with extensive experience in management, two practicing pilots and the author. More than 125 years of piloting experience at the Panama Canal was represented. Following the trip to the U.S., the group submitted a report to the Commission which not only agreed on the feasibility of simulator training for Panama

Canal Pilots, but unanimously recommended that such training be incorporated into the Canal's regular pilot training program.

Once the team decided to recommend supplementing the regular program with simulator training, it became necessary to recommend the most suitable institution. This was the more difficult decision to make. It was clear that each location visited in the United States was capable of providing adequate training although not all for the same reasons. Each place had features which argued strongly in their favor. Ultimately, Marine Safety International (MSI) in New York was recommended. In addition to the perceived quality of training, such factors as total cost, transportation, lodging, etc were considered.

MSI was selected because it was able to immediately provide simulation with a locks complex. Inasmuch as the Panama Canal has three sets of locks, this was a key factor in the final decision. A significant portion of the St. Lawrence Seaway including locks had recently been modelled for the Dominion Maritime Association in order to train DMA personnel.

Before continuing further, I believe that a brief description of the MSI simulator is in order. This system uses models of the geographic areas employed in their training scenarios. A television probe acting as the eye of the pilot is controlled, through a computer, as directed from the "bridge." The scene is then projected on a large screen which is viewed through the "wheelhouse" window. This provides a field of view of 140 degrees. Mathematical models of the selected vessels and gaming or exercise areas are incorporated. Limited vessel traffic which follow predetermined tracks is available.

COURSE DEVELOPMENT

After top management made the decision authorizing simulator training for pilots, a determination had to be made for which there appeared to be no precedent. An objective review of the training program and advancement scheme had to be made. It was decided to offer this training to pilots with approximately 18 months experience. Individuals at this level would have made approximately 200 transits of the Canal, about two thirds of which would have been as control pilot. It is at a point in their careers just before the size and tonnage of the vessels to which they will be assigned begins to increase rapidly. Subsequently, it was demonstrated that this level was about correct. As can be expected, this varied somewhat from individual to individual.

It was also necessary to determine what type of training would provide the greatest benefit to the selected pilots. As the trainees would have had limited experience on vessels over 20000 tons DWT, vessels of 30K, 50K and 80K were selected for use during simulation exercises. Certain areas of the St. Lawrence Seaway that were roughly comparable to portions of the Panama Canal were utilized initially. For the first day special exercises were planned that would acquaint the trainees with the simulator and the selected ship types as well as demonstrate certain basic maneuvers. Accelerated turns, 360 degree turns with constant RPM, crash stops, etc. were tried. The advance, transfer, tactical diameter, rate of turn and other data for each maneuver was noted.

The first two groups that were sent to New York in September 1983 were observed carefully and their critiques of the course carefully evaluated. Although, there were several shortcomings and some disappointment in the program at this stage, it was felt that the potential benefits warranted continuation. From the beginning, a warm spirit of cooperation existed between MSI and the Pilot Training Unit. This spirit continued as the course of instruction was repeatedly modified and improved following each training cycle.

When the program was first initiated, each group of trainees consisted of four individuals. During the training, they took turns as pilot-in-charge and rotated to other positions on the bridge. At times the pilots complained of boredom and fatigue. They stated that by the time the fourth man was "piloting", the first had been standing around with very little to do. It also had been noted during several training cycles that by the fourth day interest began to wane. To combat this, the lengthier exercises were eliminated and suitable replacements were developed. Toward the end of the first year of the simulator program, a more radical change was made. The size of each group was reduced to two and the duration of the course changed to 20 hours (2 1/2 days). However, the balance of the week is taken up by the second pair of pilots. The total amount of hands on experience of each trainee remains the same and interest stays high throughout. In addition, this procedure allows the Canal officials greater flexibility in matching individuals for greater compatibility.

Although the Seaway and other selected areas were acceptable, the trainees did not relate to them as it was hoped. Fortunately, MSI had, on its own, undertaken to make a model of Gatun Locks (of the Panama Canal) which was then superimposed on an appropriate Seaway's channel. This added realism and excitement to the course. In each subsequent session changes to the models and to the exercises were made to upgrade the quality of the training. Earlier, berthing exercise were conducted on the simulator using the oil berths in Valdez, Alaska. These exercises were not met with a great deal of enthusiasm even when the heading of the approach channel and the distances simulated certain areas within the Canal. When a crude model of Cristobal Harbor was substituted, interest increased dramatically and, with it, the training value.

At various times, the trainee was afforded the benefit of a bowthruster or one or two tugboats, depending on the purpose of the exercise. The vessels alternated between steam and motor ships. With the motor vessel, the number of air starts was limited. As the ships at the Panama Canal are scheduled at the locks, timing may be a factor in an exercise. Speed control, economy of maneuvers and emergency response is stressed.

THE MICRO-SIMULATOR

In the report previously cited, the evaluation team recommended to the PCC "that the current training program be augmented by acquisition of a computer-aided (desk top) training device." As a result of this recommendation, the Canal's purchasing branch began soliciting bids for a Partial Task Training Aid Marine Simulator from qualified manufacturers. In order to eliminate poorly qualified bidders, the terms of the contract required validation of the mathematical modeling. Ultimately, the micro-simulator was provided by Tracor Hydronautics, Inc. This simulator was informally name Trasim by the training staff.

The system consists of an instructor's station and a student's station. The student's station is isolated in a small room adjoining the main station. Each station has a plan view display, a bridge instrument display and two-way communications. In addition, the instructor's location has the ship control unit, the simulator control terminal and two key boards. The system also supports a printer and a color graphics plotter.

During normal training operations an instructor/operator will set up the exercise. Each station will be manned by a trainee. The one at the student's console will control the maneuvers by giving orders verbally as he would aboard ship. The other trainee will respond to these orders through the ship control unit. At times the instructor will gather other trainees at his station and will instruct by a critique of the ongoing exercise. Or he may run a prerecorded exercise for the purposes of demonstrating a particular point he wishes to make. Also, a student may be allowed to be on his own at the instructor's station, controlling the maneuvers directly with the ship control unit. This method should only be used for a specific purpose as there is a strong tendency to overcontrol The individual attempts to control the "ship" as if it was a motor boat. This tendency has been observed even with experienced pilots.

The regular training exercises on Trasim mirror those performed on the full task simulator. The employment of Trasim is not considered a substitute for the shiphandling simulator program but it has been found to be an excellent complement to it. In fact, a session on Trasim is now scheduled for everyone prior to going to Marine Safety in New York.

There has been some criticism of the use of the plan or bird's eye view and while it is recognized that the panoramic view has some advantages, the plan view is capable of providing very fine training. This view has been the traditional method used in classroom instruction, in our textbooks, on the chalkboard and even in our maneuvering diagrams. In my view, to provide an adequate field of view in a training device incorporating the panoramic view would significantly increase its cost.

The applications of this type simulator are as follows:

- To demonstrate theory or specific maneuvers
- To develop strategy
- To practice the developed strategy or specific maneuvers

As the advantages of a micro-simulator such as Trasim are somewhat obvious, they will be listed below without comment:

- Relatively low cost
- Minimal housing problems
- Can be conveniently sited
- High degree of portability

BRIEFING AND DEBRIEFING

Prior to each exercise, a thorough briefing is essential. During the briefing, the trainees are given a general description of the exercise, they are informed of the conditions that will exist at the start of the exercise and, above all, the object of the exercise is made clear. To prevent any surprises that could have a negative impact, they are honestly made aware of any limitations of the simulator that could affect the decisions of the trainee. I do not refer to those surprises that may be occasioned by reasonable shipboard emergencies.

After each trainee "pilots" the ship, a critique or debriefing is held. The plot and printout is reviewed and analysed. Constructive criticism is offered. Tact is used to avoid embarassing the student if his performance was unsatisfactory.

A week or so before leaving for MSI, the pilots return to the training center for two days of instruction. These sessions are primarily a review of the theory and fundamentals of shiphandling and include lectures, the viewing of videotapes and discussion. A lengthy workout on Trasim is part of these sessions. Some of the preliminary exercises that were normally performed the first morning at MSI are now accomplished on Trasim, thereby allowing more time for additional exercises on the big simulator. During these two days, the MSI simulator is described and all the arrangements for transportation, meals, lodging, etc. are explained.

DISCUSSION

Because of the nature of the Canal, Panama Canal pilots rely heavily on visual clues during an act of pilotage. It can be reasonably assumed that this applies to certain other pilotage districts as well. This fact has to taken into account when preparing a scenario for the simulator. This also suggests the need for an enlarged field of view for pilot training. The use of conventional navigation methods such as bearings and position fixing devices play no significant role during the pilotage. Radar is mainly employed during periods of reduced visibility but this use is limited in the heavy tropical showers that are a daily occurrence during the rainy season.

On the simulator, the need to provide adequate visual clues remains high; however, increased reliance on the bridge instruments becomes necessary. It appears that the ability to judge speed and distance is, to a degree, impaired on the simulator thereby placing greater dependence on the speed log. Under certain circumstances, such as turning in the open sea, the rate of turn indicator takes on greater importance. There is, of course, no alternative to wind direction and speed indicators. At the Canal, the instruments of greatest importance are the rudder angle indicator and the propeller revolution indicator or pitch indicator, in the case of a controllable pitch propeller. On the simulator these instruments retain their importance

It is axiomatic that people who do not wish to learn cannot be taught. Pilots reluctance to accept the benefits that can be imparted by simulator training probably fall into that category. This attitude presents a significant challenge to those responsible for their training. All exercises must have a clear objective, they must be relevant and they must be challenging. Every effort must be made to prevent any negative factors from entering into the training. Skeptics have been converted into believers when offered a sound program.

THE FUTURE

It seems that the continuation of simulator training for Panama Canal pilots is assured, at least, for the foreseeable future. At the present time, the PCC is having selected areas of the waterway specially modeled by Marine Safety as part of its ongoing program. It is expected that these special models will enhance the training by increasing the realism. This increased realism will enable the trainees to relate better with the training exercises. In addition, revised exercises will more nearly reflect the actual conditions encountered in the Canal.

With their own micro-simulator so conveniently located, the PCC can gradually extend this aspect of training to more pilots. As more experience with Trasim is gained, the applications of this equipment can be expanded. Modificatios to facilitate training have already been considered. Conceivably, preliminary research for proposed improvements could be carried out.

The Canal authorities have had discussions regarding the role of the simulator for remedial training. This was recognized as a thorny problem, not only from the aspect of acceptance on the part of the pilots, but in the difficulty in developing a program of remediation. Obviously, every individual who may require such training does not need it for the same reasons. As some pilots with apparent weaknesses do not accept that fact, the effectiveness of remedial training for that person can be questioned. Scheduling an individual into a training cycle presents an additional problem, as it would seem ill-advised to put him in with a group of regular trainees. The costs of a solo program would be unreasonable. In anticipation of the need for this type of program in the future, the PCC put two pilots of greater experience (3 to 4 years) through the simulator course. Although their feedback was encouraging, it has to guardedly evaluated as they were not selected at random.

CONCLUSIONS

These conclusions are based upon experience gained in the utilization of a shiphandling simulator in the training of Panama Canal pilots. It would not be unreasonable to expect that some or all of these conclusions would apply in varying degrees to the training of individuals from other pilotage districts. In evaluating these conclusions, it should be recalled that the pilots receiving this training already had approximately 18 months experience piloting at the Canal.

This experience has convinced me that the training of pilots by this means is feasible and, in some instances, highly desirable. The traditional method, however, cannot be abandoned as pilots must, ultimately, learn and practice their skills on the bridge of a real ship. The simulator intensifies the pilot's real world experience. In other words, it eliminates from the training equation those segments of his pilotage in which the learning experience approaches zero and concentrates on those in which the learning experience is high. It also allows him to practice difficult maneuvers in safety and trains him to respond to unusual or emergency situations.

It would be premature at this point to suggest that time spent on a simulator should be substituted for actual practical experience. At this time, at least, it would seem a difficult task to write an equitable formula that could be applied to pilotage even within one country. This is due to the great variations in the nature of the pilotage from district to district and to the different skills required in each of them. As some of the Canal pilots who were offered simulator training were at the stage when their advancement was predicated upon the number of transits completed, they were credited with two transits during the time when the program was of a week's duration. Under normal conditions, they would have made three or possibly four transits during their absence from the Canal.

It is safe to conclude, I believe, that a suitable micro-simulator can be a valuable tool in the training of pilots and other mariners. This view is supported in the literature. It is important to recognize, however, that the quality of instruction with this, as with any simulator, is a vital factor. While the micro-simulator cannot take the place of a full task simulator, it can be an excellent complement to its big brother.

The potential user of the simulator for this purpose has to evaluate many factors. High among them is the cost effectiveness of this training which is difficult to measure, especially in the short term. In this respect, the FCC believes that if one serious accident is averted because of the improved quality of training, the cost of the training will have been defrayed.

Even when funding is available, there are factors that must be considered as they can significantly influence the value of this type training. Some of these factors have been listed below:

- The value of the simulator training will vary with the nature of the pilotage. In certain districts, the benefits will be marginal while in others they will be large.

- The objectives of the program, in general, and each exercise, in particular, must be clearly defined.

- The simulator at the training facility must be capable of satisfying these objectives.

- The course of instruction must be tailor made to suit the specific requirements of the pilotage and must be designed to match the experience level of the pilots being trained.

- The scenarios should be challenging and should be no longer than necessary to meet the objectives.

- The curriculum must be relevant.

- A thorough briefing before and a critique after each exercise is essential. The briefing should include a review of the objective(s). The critique should be comprehensive and done in a manner so as avoid discouraging or embarassing the trainee.

- The type and capability of the simulator should be taken into account.

As long as the simulator is capable of satisfying the training objectives, the type of simulator is not of major importance. To illustrate, a nocturnal simulator, using a spot projector only, might be perfectly suited for training pilots in a selected port while it would be totally unsuited for the Panama Canal pilots. Once having met the training objectives, therefore, it cannot be stated that one type of simulator should be considered to the exclusion of other types. Considering the needs of the PCC, a list of the features required of a simulator is provided below:

1. The visual scene
   - A minimum field of view of 180 degrees
   - A field of view of view over the stern of 30/40 degrees *
   - Textured sea surface

2. The wheelhouse
   - Standard bridge instruments and equipment (position fixing aids not required)
   - Simple radar (no plotting)
   - Speed log, fore and aft/lateral
   - Wind velocity and direction indicator
   - Internal and external communications

3. The environment
   - Current/tide
   - Wind/gusts
   - Sounds and vibrations *

4. The interactive forces
   - Shallow water effect
   - Bank effect
   - Intership action

5. Own ship
   - At least four of various types and tonnage
   - Turbine and motor ships
   - Bow thruster *
   - Twin screw *

6. Other ships
   - More than two
   - Controllable by instructor during exercise
   - Tugboat forces (2)

Items marked (*) are deemed desirable but not essential.

At this time, the Panama Canal Commission appears to be pleased with the direction that the pilot training program is moving. Through the shiphandling simulator course a new dimension has been added to what has been a successful training program and it is envisioned that Trasim will provide the means of bonding the shipboard, theoretical and simulator facets of this program together into a homogeneous unit that will ultimately result in a pilot of the highest caliber.

REFERENCES

Ankudinov, V. and Jakobsen, B. (1984), Theoretical Basis for Partial Task Training Simulator, Technical Report No. 84022.0124-2, Tracor Hydronautics, Inc.

Hammell, T.J., and Gardenier, J.S., (1984), Bridge simulator training system differences: third mate - pilot, Proceedings MARSIM '84

Panama Canal Commission, (1982), A Report on the Issue of Simulator Training for Pilots, unpublished.

Perdok, J and Elzinga, Th., (1984), The application of microsimulators in port design and shiphandling training courses, Proceedings MARSIM '84.

Micro-Computerized Maritime Traffic Simulator

T. DEGRE (I R T)
X LEFEVRE (O R I O N)

I R T : 2 Avenue du Général Malleret-Joinville
94114 ARCUEIL CEDEX, FRANCE

O R I O N : 8 Rue Saint Marc
75002 - PARIS, FRANCE

Summary

This paper describes a micro-computerized maritime traffic simulator, which reproduces the behaviour of a given traffic flow display of most modern Vessel Traffic Services.

This tool can be used in the three following subjects :

- study of traffic organization in a given sea area,
- testing new solutions as far as surveillance and information to navigation are concerned,
- training for VTS radar operators.

In 1976, O.R.I.O.N. began development work for the Transport Research Institute (I.R.T.) and the Lighthouse and Beacon Department, concerning a computerized simulation model for maritime traffic in the Pas-de-Calais area, to examine the possibility of improving the present organization of ships' traffic, so as to increase safety in this area. In 1978, after modifications to the traffic separation systems of Ouessant and Casquets, and as a result of the Amoco-Cadiz catastrophe, it was decided to extend the simulation model to the Channel as a whole, so as to study proposals made in France and the United Kingdom to modify traffic organization in this area.

At present, the I.R.T. uses a simulation model known as the "M.A.R.I.N. Model" (Computerized analysis and research model for navigation), which is used to simulate traffic in the Channel, and to evaluate the risks of collisions between ships as a function of a certain number of criteria.

From the results obtained, it is clear that this model has played an essential role in the definition of the traffic patterns in the Channel, proposed by France and the United Kingdom and adapted by the O.M.I. (O.M.C.I.) in 1981.

More recently, in January 1982, the M.A.R.I.N. model was used to study the effects of a bridge across the Channel on navigation safety.

Moreover, most of this model was used by the Liverpool Polytechnic maritime research team, to study traffic in the Irish Sea.

A simulation model of this type offers numerous possibilities which have not yet been explored. It is for this reason that it has been decided to transcribe the model to a micro-computer, so as to provide a tool enabling:

- a rapid preliminary study of traffic organization in a given location
- simulation of a synthetic image of a surveillance radar (coastal or harbor), and therefore assistance in choosing a site for installation of a station, evaluating operator workload and assisting in operator training.
- testing and development of algorithms for maritime traffic (anti-collision from a surveillance center, data bank, etc.).

This maritime traffic simulator consists of a number of programs written in high level PASCAL language, and enabling:

- display of marine charts
- ships' traffic generation
- definition of the organization of the traffic studied
- ships' movements.

The purpose of this micro-computerized maritime traffic simulator, constructed from the M.A.R.I.N. model, is to present a given traffic flow in a given sea area, allowing for a fixed organization diagram, the resulting product being presented in the form of a real time image displayed on the TV screen associated with the micro-computer. The operator is provided with a certain number of controls enabling him to control this image (off-center, scale change, ships' code display, etc.); this will be explained appropriately throughout the text.

1 - REVIEW OF THE M.A.R.I.N. MODEL STRUCTURE (See bibliography)

1.1. Model basic principles

According to the data which will be examined below, the model generates a variable number of ships of determined types, originations and destinations at random times, assigns a route to these and causes the ships to move at their speeds and headings, along the route to their destination.

The operation consisting in assigning a route to a ship depends on the origination/destination of the ship, its type, its considered organization diagram (traffic separation devices, associated rules, beaconage), and the behavior of the man/ship system.

The manoeuvres performed by ships to avoid collisions are not taken into account at this stage. Thus, the only ships' manoeuvres which are simulated at this level are successive heading changes which ships must make to reach their destination and, insofar as possible, accomplish "clean" navigation (avoid natural obstacles, avoid contraventions to regulations associated with proposed traffic separation systems).

1.2. Model data and hypotheses

The data and hypotheses used in the model concern: the area marine chart, the characteristics of traffic through the area for a given period, the characteristics of each of the organization diagrams considered and the behavior of navigators.

a) The area marine chart

The marine chart used in the model comprises: the coastal contour, an outline of the traffic separation system or systems, corresponding to the organization diagram studied, beaconage (buoys, light ships, light buoys, etc.) and the names of ports, capes or landmarks.

b) Traffic

From enquiries concerning traffic movements in the area studied, it is possible to determine and enter the following in the model:

- the number, types and originations/destinations of ships frequenting the area during a given period.
- the time laws for entry of ships into the area considered. These may be random or deterministic (e.g. ferry schedules)
- laws concerning ship navigation speeds.

c) Organization diagram characteristics

The data of an organization diagram comprises:

- a geographic definition of the traffic separation systems
- definition of special regulations associated with these systems
- recommendation of a navigation aid system (characteristics and geographical locations).

d) Navigator behavior

In addition to its origin/destination, the actual course which a ship follows depends on its type and the organization diagram considered, human factors concerning choice of this course and ship's navigation aids, used to keep to this course.

It was found necessary to state the following hypotheses:

- Course choice

All ships respect the particular regulations associated with traffic separation systems, together with regulation 10 of the international rules concerning traffic separation systems.

This implies that longitudinal traffic (traffic using the systems) navigates within the navigation corridors defined by these systems, and that crossing traffic (traffic cutting across the systems) does so at an angle which is as near as possible to a right angle.

It sometimes occurs that ships are in turn considered as belonging to longitudinal traffic or crossing traffic; for example, this applies to ships desiring access to a port (or leaving a port) or leaving (or entering) the system.

For this type of ship, we use the hypothesis that it chooses the shortest course compatible with regulations and geographical and nautical constraints.

Finally, in most locations to which traffic separation system rules apply, there are coastly navigation areas, in which navigation can be two-way. In such cases, we have also used the hypothesis that ships choose the shortest route, in terms of geographical and nautical constraints.

These hypotheses result from common sense considerations. It has been possible to verify that these give results very close to reality for areas for which surveillance radar screen photographs provide accurate knowledge of the paths actually followed by ships (Jobourg and Gris-Nez).

- Course holding

It is considered that ships have adequate on-board facilities, in correct operational condition, so as to follow and check the courses selected. However, to allow for yaw (lurching), drift due to wind and current and localization accuracy, it was considered that there was a certain dispersion on the course actually followed by the ships with respect to the desired course.

1.3. Simulation mechanism

Having stated the hypotheses of navigator behavior, it is then possible to associate a navigation "path" for each type of origin/destination.

These paths consist of a corridor, inside which all ships of this origin/destination will navigate.

The navigation corridor, corresponding to a given path, is constructed using a certain number of heading change lines (the first is known as the entry line, and the last as the exit line).

To recreate dispersion which occurs in reality, statistical entry laws are used which enable the ship to arrive on the corridor entry line, and transition matrices, which enable the ship to change from one heading change line to another, until it exits from the area.

2 - PRESENTATION OF THE TRAFFIC SIMULATOR

2.1. Introduction

In order, a brief description of:

- the various data files
- various programs
- operator-simulator dialogue

is given below.

2.2. List of data files

The data files required for simulation can be broken down into three categories:

- The MARINE CHART files:

These files, which are used to define the marine chart, are as follows:

- origin, mean latitude and initial scale of the chart (INIT)
- island and coastal contours (COAST)
- traffic separation systems (DISP)
- main beaconage (BALI)
- names of ports or landmarks (capes, islands, major navigation aids, etc.) (NOM).

- NAVIGATION files:

These files, which are used to define navigation organization in the area and diagrams studied, are as follows:

- navigation corridors corresponding to various origins/destinations (PARC)(abbreviation of "parcours" (path))
- ship's heading change lines (LIGN)
- statistical entry laws for geographical entry lines to the area studied (LOIE)
- transition matrices for changeover from one heading to the next (MATT)

- TRAFFIC file:

This file contains the arrival times, number, origin/destination and speed of the various ships operating in the area and the organization diagram studied; these ships are classified according to chronological order, as a function of their arrival times.

This file is created from statistical data files concerning the number of ships on each path, the time laws, or deterministic schedules and speed distribution laws.

2.3. List of programs used

- MARINE CHART PROGRAM

This program is intended to generate digital data files enabling plotting of a marine chart for the location chosen. This program enables creation, modification and printout of these files.

- NAVIGATION PROGRAM

This program is intended to manage the digital data files enabling organization of navigation in the area and the chosen diagram to be defined.

- THE VERIFICATION PROGRAM

This program is intended to graphically define the validity of the digital data and files used to define:

- the area marine chart
- the navigation corridors corresponding to the organization diagram studied.

This, therefore, is merely a checking tool, but it is indispensable for fast and reliable compilation of files required for the traffic simulator.

- TRAFFIC PROGRAM

This program is intended to create, modify or read statistical data files concerning the traffic to be introduced into the area and over a second stage, to generate a traffic file, which, for each of the ships simulated, will contain:

- the number
- the arrival time in the area
- the path number
- speed.

Four statistical data files are used:

- the STAT file, which contains the number of ships operating on a given path per day
- the LOIH file, which contains the various arrival time laws
- file HORA, which contains the passenger ship schedules
- file LOIV, which contains the speed distribution laws.

- SIMULATOR PROGRAM

This program is intended to display traffic circulation within the area and organization diagram chosen, using the above-mentioned data files.

The display is made with or without a radar screen (see examples below), as desired. The user initially defines the simulation time period (minimum 1 minute) and the simulation duration.

A detailed description of operator-simulator dialogue is given in the next paragraph (see figure 1).

2.4. Description of operator-simulator dialogue - Examples

As can be seen from these examples, the dialogue commands are displayed permanently on the lefthand table of the screen. In the righthand table, the elapsed time is permanently displayed as a function of the times previously selected, together with the present scale. Moreover, a table is provided to display the characteristics of one or more ships, as designated by the operator. These commands are implemented from the microcomputer keyboard keys.

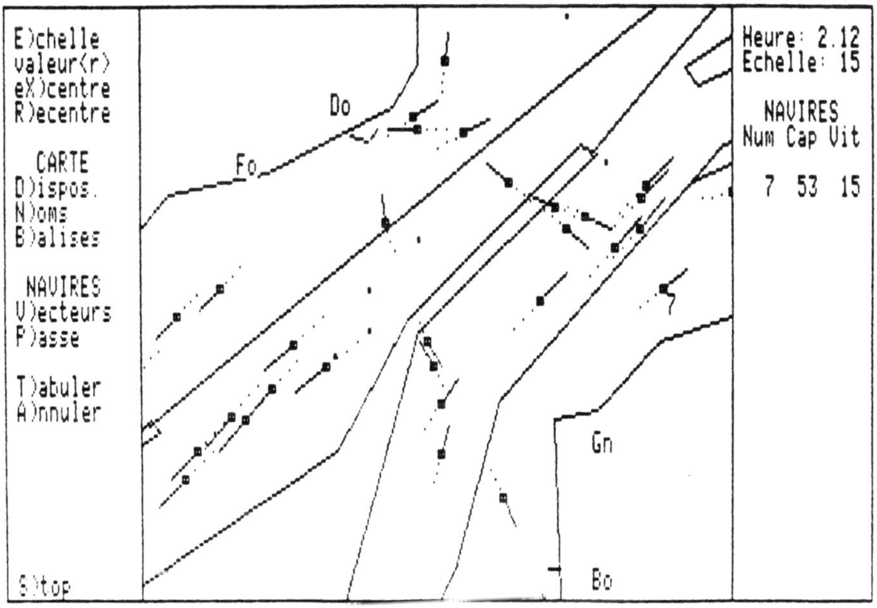

Figure 1

EXAMPLE OF TRACK DISPLAYS WITH PREVIOUS POSITIONS AND SPEED VECTORS

The track speed vectors represent an extrapolation of their course and speed over a given period of time (in this case, 6 minutes).

Speed vectors provide initial data to be used in judging the risks of collision between one or more ships.

The following controls are provided:

- General controls:

  - Scale change, operated by key E and insertion, as integers, of the new scale value desired (minimum: 1 mile, maximum: 500 miles)

  - Image off-centering: When the operator presses key X, he can then move a cross along the four directions, North, South, East and West, using the arrow keys on the keyboard. The final position of this cross then designates the new center of the next image;

  - Image recentering: Via key R, the operator can reset the image to the initial center, while conserving the present scale.

- MARINE CHART display controls:

  The operator can display or remove the following:

  - traffic separation systems (key D)
  - landmark names (key N)
  - principal beaconage (key B).

- Ships' echo display controls
  (refer to examples provided for further details)

The operator can display or remove:

- the speed vector (heading and speed) for all echoes displayed (key V)

- previous positions for all echoes displayed (key P).

The operator can also designate one (or more) ships so that their characteristics are displayed in the corresponding table.

Having pressed key T (for Tabulate), the operator can then move a symbol (small circle) using the arrowed keys of the keyboard, up to the present position of the ship's echo for which he wishes to obtain the characteristics. Movement of this symbol can be made at high or low speed. To cancel the ship's echo characteristic display, the same operations are repeated, but key A (for Cancel) is pressed first.

- End of simulation control:

The operator can stop simulation instantaneously by pressing key S.

2.5. Overall structure

A simulation session is based on preliminary creation of three file categories:

- MARINE CHART files, created or modified by means of the MARINE CHART program,

- NAVIGATION ORGANIZATION files, created or modified by means of the NAVIGATION profram.

- Both file categories are checked using the CHECKING file.

- The TRAFFIC file is created using a program of the same name and a statistical file, concerning the number of ships on each path, the arrival time laws and the speed laws.

When these files have been created and checked, and after modification, the simulator program is ready to be run.

If the radar screen display option is chosen, it is necessary to use the radar screen program to create a file of the same name, which will be used by the simulator program.

The SIMULATOR program then exploits all these files to create a dynamic image displayed on the TV monitor connected to the computer.

During simulation, the operator intervenes at two levels:

- during initialization, to define the area to be simulated, the time period, the simulation duration and display or not of a radar screen background,

- on execution, where he can control the marine chart and/or ships' echoes display.

CONCLUSION

Throughout the description of this maritime traffic simulator, an example concerning the surveillance station at GRIS-NEZ was used.

It will be clear that, from the moment in which the digital data required are available, it is possible to use this simulator for any other area.

In editing programs, all efforts have been made to ensure that the operator machine dialogues are accessible to persons not acquainted with data processing. To provide an order of magnitude, the time required to prepare the Pas-de-Calais simulation (in which the digital data to be entered are numerous) was approximately three days.

It should be noted that, in its present state, the simulator can provide assistance:

- in traffic organization prestudies for a given area

- in selecting a site for installation of a radar surveillance station

- in evaluating operator workload in surveillance center

- in training operators.

At present, O.R.I.O.N. is using this simulator to study and concretely test algorithms or display modes enabling anti-collision control from a surveillance center.

As an example of this, an operator could be provided with the possibility of displaying the surroundings of a ship designated either by the operator himself or subsequent to an automatic alert.

Presentation of these surroundings could be based on the type of display available on the screens of the ARPA radars at present fitted to some ships.

This research is not intended to supervise the anti-collision tasks of navigators, but rather, for example, to provide assistance in the event of equipment technical failures.

The interest of a micro-computer simulator in this field is its independence, transportability and low cost.

In fact, it is possible to develop, test and refine anti-collision algorithms without disturbing operator tasks at existing centers, and also to offer them an on-site capability of changing situations as a function of their remarks and criticism.

At present, the micro-computer used is an APPLE III (256 Kbytes of main storage). However, all programs offered can easily be transposed to other micro-computers of sufficient memory capacity using the PASCAL language.

The Minimum Manoeuvering Range as a Criterion to Evaluate the Risk of Collision Between Ships

BERND HAMPEL

Universität Hamburg
Institut für Schiffbau

Abstract

To evaluate the risk of collision between two ships in a hazardous encounter, it is necessary to establish a criterion which quantifies the danger for the involved ships.

For this purpose the Minimum Maneuvering Range seems to be a suitable parameter. It describes the minimum distance from the other ship, which one ships needs to avoid a collision, when starting her evasive maneuver. To calculate the Minimum Maneuvering Range and the corresponding evasive maneuver the
- actual maneuverability of the own ship and
- possible maneuvers of the other ship

have to be taken into account.

Two methods to solve this problem are considered.

The results for several encounters of two ships with variable rudder application are computed. A simple and definite representation of the necessary maneuvering range and actions on the radar display is suggested.

Nomenclature

Abbreviations

A	Another ship
CPA	Closest point of approach
H	Hull
IRPCS	International Regulations for Preventing Collisions at Sea
MMR	Minimum Maneuvering Range
MR	Maneuvering Range
MPD	Minimum Passing Distance
O	Own ship
P	Propeller
R	Rudder
S	Seaway (e.g. wind, waves, shallow water, etc)

Symbols

f	Vector function
I_z	Moment of inertia about z-axis
j	Index

m	Mass
N	Moment about z-axis
r	Range
W	Value function
X,Y	Forces along x-, y-axes
$\underline{x}$	Nondimensional coordinate
x_G	Coordinate of center of gravity
u,v	Velocity component along x-, y-axes
ψ	Heading angle
ϕ_O, ϕ_A	Control variable

1 Introduction

1.1 General

Usually there are no problems in stating the division of responsibilities between two approaching vessels. But it is more difficult to choose the proper and most effective action to avoid the collision. For this purpose the navigator has to regard all circumstances of the actual situation. The decision on board, as to which maneuver is to be taken at what time, is generally made in relation to

- the legal position in accordance with the International Regulations for Preventing Collisions at Sea (IRPCS), and
- the threat, presented by the maneuver actions of the other ship and/or the supposed plotted or calculated passing distance.

Such a finding can be wrong, even if it was made by an experienced seaman and with the necessary precaution.

The 'decision-making-process', as sketched in Fig. 1, during an approach is always made on the basis of scanty information (expresses by dotted lines in the Fig.). Usually the maneuverability of the other ship and her maneuver and/or maneuver intentions are unknown. Very often one also has only little knowledge about the behavior of own ship in extreme maneuvers due to the influence of wind, wave, draft effects etc. in the actual seaway.

In relation to these aspects it seems to be very useful for the navigator to have a procedure available which on the one hand characterizes a dangerous situation and on the other determines a maneuver suggestion which is in accordance with the IRPCS.

To solve this problem in a satisfactory manner, the maneuverability of the own ship and the uncertainties concerning the other ship's maneuver intentions and her characteristics have to be taken into account.

1.2 Maneuvers

In this paper a maneuver is characterized as a continuous change of heading into one direction due to steady rudder application. Continuing straight on after a course change counts as a half maneuver.

A maneuver in this paper consists of two elements:
- the maneuverability -- technical element
- the maneuver 'intention' -- managing element

The maneuverability has definite values for each ship which may change with the conditions of the seaway, draft, etc. In connection with the rudder angle it determines the dynamic behavior of the ship. The simplest description can be made by the values for the minimum turning radius and forward speed during turning.

The maneuver intention is an output of the decision-making process. It influences e.g. the application of the rudder. Its value varies between -1, 0 and +1 (pessimal, neutral, optimal). In this context, pessimal doesn't simply mean the contrary rudder angle to the optimal, but rather in the true sense of the word, it expresses the behavior of the ship with respect to the goal - avoidance of collision.

If a maneuver tends to maximize the final passing distance, it is called an optimal maneuver. When one ship acts due to error, misinterpretation or failure of equipment in a manner which would minimize the distance at CPA, she acts pessimally. One who doesn't react at all remains neutral.

The maneuver intention affects the result of the maneuver a lot more than the maneuverability.

1.3 The Minimum Maneuvering Range

Beginning with the fact, that the distance at the moment one ship starts her initial maneuver to pass another at a safe minimum distance (MPD: Minimum Passing Distance), can be calculated, if the executions of the maneuvers and the maneuver characteristics of both ships are known, the whole evading operation is reduced to its technical base. The value of MPD is set by each of the navigators for himself and will differ with encounter geometry and traffic situation.

The variation of all possible maneuver combinations yields a whole set of distances. Those distances, which one gets with the maneuver combination
- own ship: action according to IRPCS
- other ship: arbitrary maneuvers,

are named Maneuvering Range (MR).

Obviously there are no rational reasons for any hypothesis that presumes a definite behavior of an approaching ship in the risk of collision (e.g. standing on of the burdened ship in situations covered by Rule 17 a ii.)

Whether she is burdened or not, recognizing the deadly danger, she will try to evade. Whether such a reaction will help to avoid the collision or, in connection with the maneuvers of the opposite ship cause a collision, is hard to answer.

The definition of the Minimum Maneuvering Range (MMR) is: The MMR is that MR which the vessel acting according to the IRPCS needs to make a safe passage with the other ship on the assumption that her maneuvers are all pessimal.

It will be understandable that the MMRs with the pessimal behavior of the opponent ship are those distances with the greatest values. But using only these MMRs as the final definition one is independent of the arbitrary reaction of the other ship.

It should be mentioned that it is maybe more sensible to use the time which both ships need to sail with constant speed and course to the supposed CPA instead of the corresponding MMR. Such a time span is almost independent of course and speed of the opponent in the current encounter, as opposed to the distance which varies considerably (see /5/).

2 Mathematical models

2.1 Differential Game Model

The analytical basis of the theory of Differential Games as formulated by Isaacs /1/ is used by Merz /2/, Miloh and Sharma (/3/,/4/,/5/) for the determination of optimal evasive maneuvers. Here these models are applied.

The dynamic system comprising the two approaching ships can be expressed with the kinematic equation

$$\dot{\underline{x}} = f(\underline{x}, \phi_O, \phi_A) \tag{1}$$

where $\dot{\underline{x}}$ describes the motion of the ships in relative coordinates and $\underline{x}$ is the vector of state variables comprising the position and the course angle.
ϕ_O, ϕ_A are control variables for own ship and the other ship. They are at the volition of the navigators and can be interpreted as normalized rates of turn.
The function f connects the state and control variables.

The maneuverability of each ship is taken into consideration by the speed and the minimum turning radius.

In the game the closest approach characterized by

$$\dot{r} = 0 \tag{2}$$

is the terminal condition, where r stands for the range at CPA. In our game it is identical with the MPD, when action takes place in maneuver zone II, or the body-contour, when trying to evade in maneuver zone III (see 2.3.2).
Up to now this model only uses ships with circular body-contours, but in /6/ and /7/ the basics are developed to extend it to ships whose contours are approximated by ellipses or straight line segments.

According to Isaacs the optimal game is governed by the first-order partial differential equation of the Hamilton-Jacobi-type

$$M\left(\sum_j W_{x_j} f_j(\underline{x}, \phi_O, \phi_A)\right) = 0 \tag{3}$$

The operator M reflects the players objectives. If ship O has optimal and ship A has pessimal maneuver intentions, then

$$M = \max_{\phi_O} \min_{\phi_A} \tag{4}$$

W is the value function. It describes the attainable passing distance in the game.

Equation (3) can be solved by integrating its characteristic equations.

Besides the MR the associated helm orders and times are results of the calculation.

2.2 The Iterative Model

In this model the basics are the equations of motion for each ship with

$$m\ (\dot{u} - v\dot{\psi} - x_G\ \dot{\psi}^2) = X = X_H + X_R + X_P + X_S \tag{5}$$

$$m\ (\dot{v} - u\dot{\psi} + x_G\ \ddot{\psi}) = Y = Y_H + Y_R + Y_P + Y_S \tag{6}$$

$$I_z\ddot{\psi} + m\,x_G\ (\dot{v} + u\dot{\psi}) = N = N_H + N_R + N_P + N_S \tag{7}$$

(For indices see Nomenclature)

They represent balance of forces and moments on a ship in a ship fixed coordinate system as described in /8/.

Starting at a point where the forces and velocities are known they can be calculated for any future time and any maneuver. Integrating the velocities one obtains the maneuver path of the ship.

The comparison of the paths of both ships, regarding the maneuver, times and e.g. MPD results in the determination of MR.

The effects of the environment can be considered in particular, also the loss of speed and revolutions during the maneuver.

A similar approach was first used in /9/ to calculate maneuvering ranges for combined rudder and engine maneuvers.

2.3 Adaption of the models

2.3.1 Number of maneuvers

In the model developed in 1.3 the MMR becomes infinite if the other ship is faster and her maneuverability exceeds a critical amount in relation to own ship.

This fact signifies: the faster ship can always collide with the slower one, if the maneuver time and the number of maneuvers are not limited.

In practice the number of possible pessimal maneuvers of the ship are restricted in general to zero, one, one and a half or two and dependent on the state of approach and the number of maneuvers of own ship.

The number two as the maximum limit is generally deducible from the IRPCS. In /10/ the suggested actions for own ship can only be effective, if one supposes that the opponent acts pessimally only twice.

But there is another, simpler argument for two pessimal maneuvers - the remaining time. Usually the approach is so fast in respect to the maneuverability of the ships, that there are only one or two actions possible.

In the mathematical models introduced in 2.1 and 2.2 the own ship should be able to perform two and a half maneuvers, if required, to always arrive at a realistic solution for MMR.

2.3.2 Maneuver zones and the adjustment of the MMR

To match the maneuver suggestions within the limits of the IRPCS, it is necessary to divide the area around a ship symbolically into zones with different legal dispositions of responsibilities. This is done in Fig. 2.

Outside of border I in Fig. 2 no evading is necessary because the distance at CPA will be so big, that neither vessel, or at least not the stand-on vessel feels threatened by the other ship.

Between border I and II is maneuver zone I. This is the part where the evading movement usually takes place. If both ships continue their original courses, the passage will occur at a distance where the burdened ship feels endangered. Here the burdened ship maneuvers.

Crossing border II the clearance at CPA (e.g. MPD) will be less than acceptable, the risk of collision is there. This is just the line which is called MMR in 1.3. If the opponent crosses this line, own ship will no more be able to keep her out of the MPD in CPA if ship A acts completely pessimally.

At border III the cooperation of both ships is necessary for them to pass each other without colliding.

The coordination of border and number of maneuvers as used in 2. are shown for own ship and the other ship in Fig. 3.

2.3.3 Selection of the MR

Each of the models compute the MR for a definite encounter of two ships. It depends, besides environmental effects, on the initial course angle, the speed ratio, the position of ships and the maneuvers against one another.

In practice the position of the other ship cannot be measured by radar with the necessary accuracy (see /11/). Therefore it seems to be advisable to select the largest possible MR, as demonstrated in Fig. 4, from all the local maneuver distances.

Fig. 4 shows the required maneuver distances, calculated with the Differential Game Model for a given initial course angle, minimum turning radius and speed ratio, when only ship O maneuvers and ship A stands on.

Both ships are pictured as circles with the radius equal to half length.

The solid line stands for optimal maneuver distances attained with optimal maneuvers. Those maneuvers yield the absolutely shortest maneuver distances to pass the opponent in MPD, but require a great accuracy to determine the position of A when reaching the line.

The dotted line reflects maneuver distances when O acts according to Rule 17 a ii of the IRPCS (only starboard-turns permissible). The MR, as defined in 1.3, measures the designated maximum distance.

The bearing where the optimal rudder application changes from port to starboard depends on the speed ratio and the minimum turning circles.

If, at constant speed ratio, the course angle in a two ship encounter is varied, the direction of approach (compass bearing) changes. The plot of MR, which cover the bearings of the own ship is named global MR (see also Fig. 5 and the accompanying explanations). Those plots give an impression of the danger of collision in relation to the direction of encounter.

3 Results

3.1 Review of the present stage of the development

Concerning the application on board the models considered are able to lower the risk of collision in two ways:

- by giving the optimal helm order for the last-minute-turn according to Rule 17 b (or 19 e);
- by determining the MMR to evaluate the risk of collision.

Because of its simplicity and character the Differential Game Model is preferable when solving conflicts dealing with the first aspect. But that aspect is not the topic of this paper.

Solutions for the second point appear to be more valuable for a carefully acting navigator, in order to avoid the risk of collision at an early stage (e.g. Rule 17 a ii or Rule 19 d).

Here both models are in principle applicable.
Up to now the MMR and the pertinent maneuvers can be determined if the other ship is slower or somehow restricted to not more than one pessimal maneuver.

One can assume that the navigators on board will choose the MPD similar to the size of the ship length. The effects of the seaway, speed reduction or drifting during turning are of the same order and therefore have to be considered. This can be done with the Iterative Model in detail, but with the Differential Game Model in the present state only on a global level.

To analyze encounters, when the other ship is the faster one, and own ship needs more than one and a half maneuvers to evade, the Differential Game Model has to be extended, and this has not been done yet.

With the Iterative Model one can calculate all possible maneuver combinations. But the amount of the maneuvers in this case increases the variation of the rudder application, which has to be done in direction, size and time, so immensely, that it seems to be more effective to search for a decision-making-model first, which would preselect the most dangerous maneuver combinations.

3.2 Global Maneuvering Range - Examples

Figure 5 to 8 are examples of the calculation of global MRs.

The vertical axis of the diagram shows the MR, the horizontal the bearing from own to the other ship. It is counted from ahead (360 degree) over the port-side (270 degree) to aft (180 degree) of Ship O which is, including the MPD, symbolically drawn in the right hand corner.

Fig. 5 - MR by variation of the speed of the other ship.
The lines are calculated for the maneuver combination
- A stands on
- O evades to hard starboard

with the Iterative Model.
The little ship-symbols on the MR-lines show in the direction of ship A's heading and the point of her body-contour, against which MR is measured.

Fig. 6 - MR due to starboard course changes of O when A stands on computed with the Iterative Model.

Fig. 7 - Comparison of the results of the two models for optimal maneuvers of O when A stands on.
Fig. 8 - Comparison of MMR and MR for different maneuver intentions of A.

3.3 Suggestion for the presentation of MMR on radar

To make correct and fast MMR determinations and maneuver suggestions the use of an adequate computer is absolutely necessary.

The required data should be automatically and continuously available from the radar and/or other facilities. The results should be displayed for a preselected object on request. A time span of less than 10 seconds (three revolutions of the radar scan) appears acceptable for computing.

The characters printed on the screen which indicate the MMR and the accompanying maneuvers have to be very simple and may not give more than three pieces of information at once, such as MMR, direction of first course change and number of expected maneuvers (see Fig. 9).

To observe special requirements - e.g. acting contrary to the rules - an alternative maneuver with opposite helm order should be computed.

The suggested maneuvers should not cause a collision with a third ship in the vicinity.

References

/1/ Isaacs, R.: Differential Games. John Wiley and Sons, New York (1965)

/2/ Merz, A.W.: Optimal Evasive Maneuvers in Maritime Collision Avoidance. Navigation 20 (1973) 144-152

/3/ Miloh, T.: Determination of Critical Maneuvers for Collision Avoidance Using the Theorie of Differential Games. Inst.f.Schiffbau, Hamburg, Ber. Nr. 298 (1974)

/4/ Miloh, T.; Sharma, S.D.: Maritime Collision Avoidance as a Differential Game. 4th Ship Control Systems Symposium, The Hague, Netherlands (1975)

/5/ Sharma, S.D.: On Ship Maneuverability and Collision Avoidance. Proc. Second WEMT Conference London (May 77) 189-200

/6/ Miloh, T.: The Game of two Elliptical Ships. Optimal Control Applications & Methods. Vol. 4, 13-29 (1983)

/7/ Olsder, G.J.; Walter, J.: A Differential Game Approach to Collision Avoidance of Ships. Twente University of Technology, Enschede, The Netherlands (1977)

/8/ Soeding, H.: Prediction of Ship Maneuvering Capabilities. Schiffstechnik 29 (1982)

/9/ Kenan, G.: Collisional Avoidance between Surface Ships at Short Range. SNAME Northern California Section Paper, Los Angeles, USA (1972)

/10/ Hilgert, H.: Die Bedeutung von Manöverkennwerten bei der Auslegung von Kollisionsverhütung für die Bedingungen der verminderten Sicht. Diss., TH Warnemünde/Wustrow (1980)

/11/ Hampel, B.: Bestimmung des Manöverabstandes für den Kurshaltepflichtigen zur Vermeidung von Nahbereichssituationen. Institut für Schiffbau, Hamburg, Bericht Nr. 434 (1983)

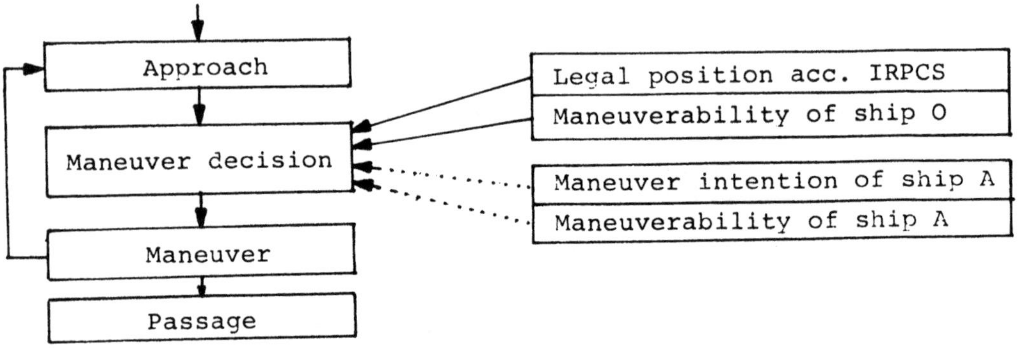

Fig. 1 Decision-making-process

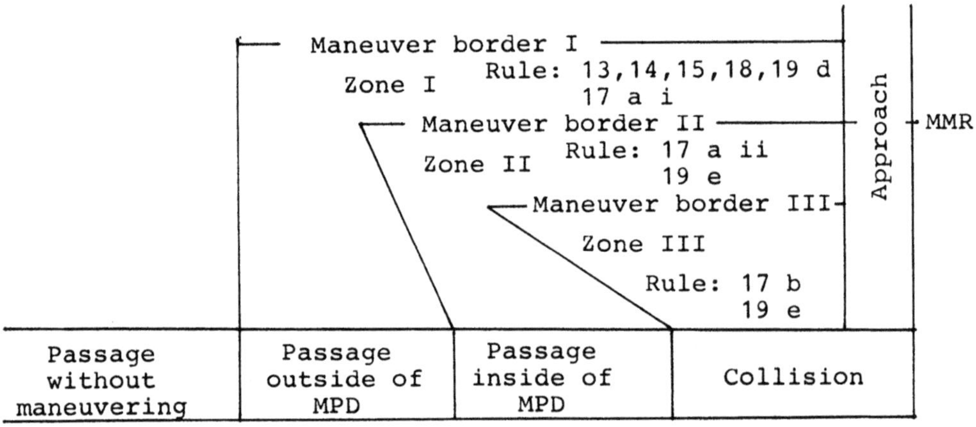

Fig. 2 Maneuver borders and the adjustment of the MMR

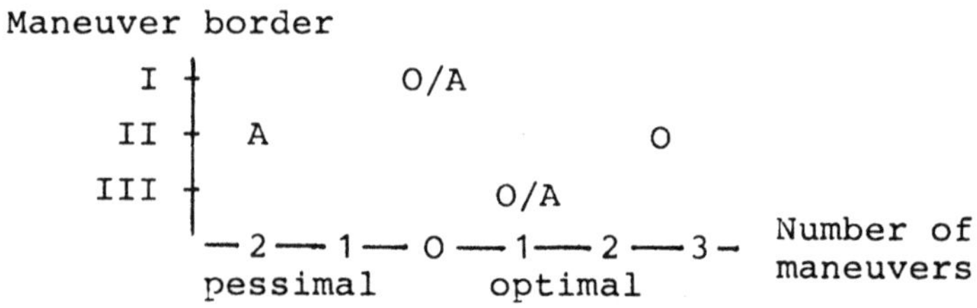

Fig. 3 Coordination of maneuver border, maneuver intention and number of maneuvers

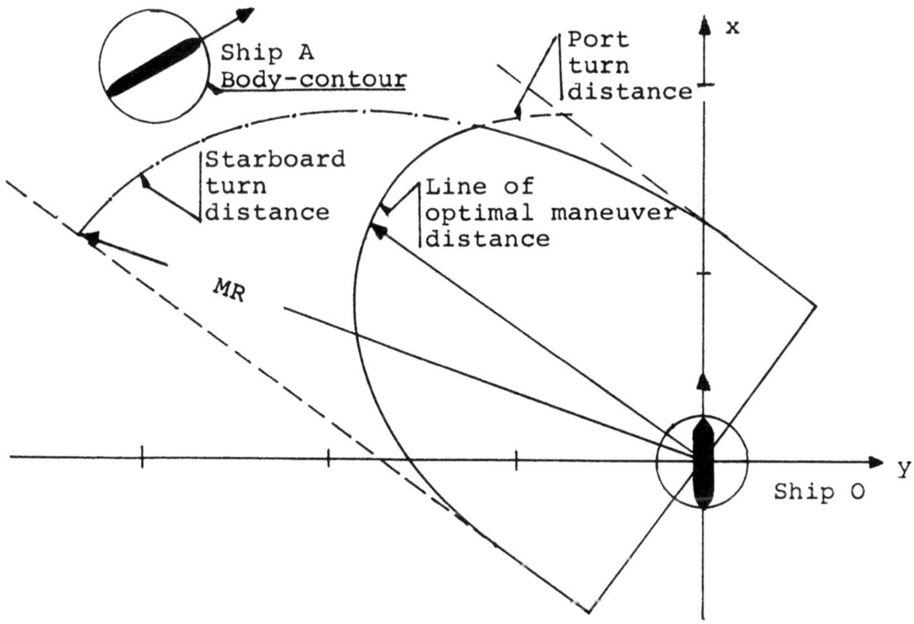

Fig. 4 Local maneuver distance

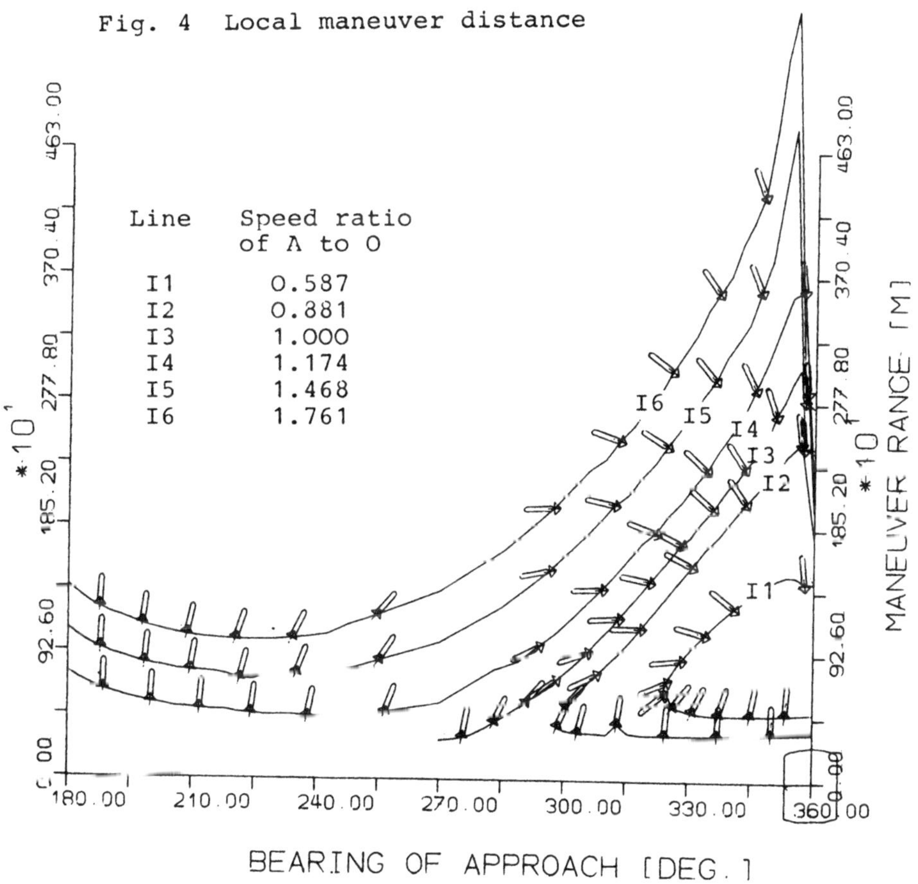

Fig. 5 MR by variation of the speed of A

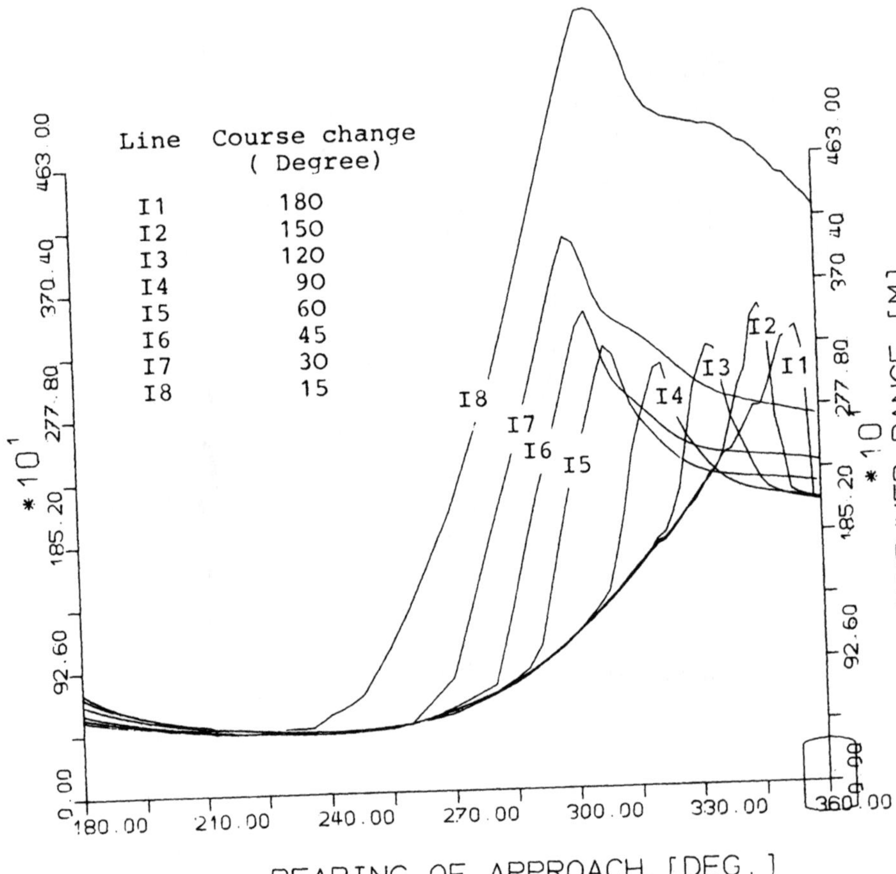

Fig. 6 MR due to starboard course change of O when A stands on.

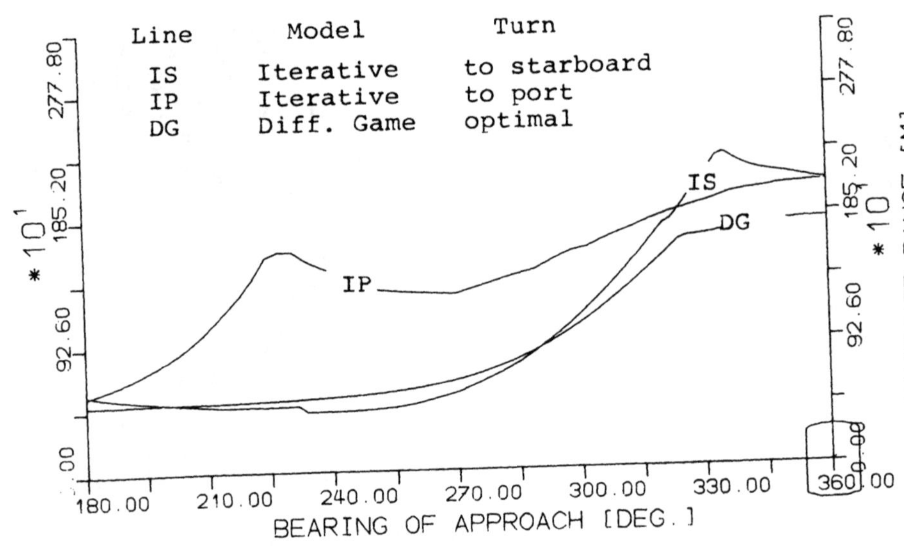

Fig. 7 Comparison of the results of the two models for maneuvers of O when A stands on.

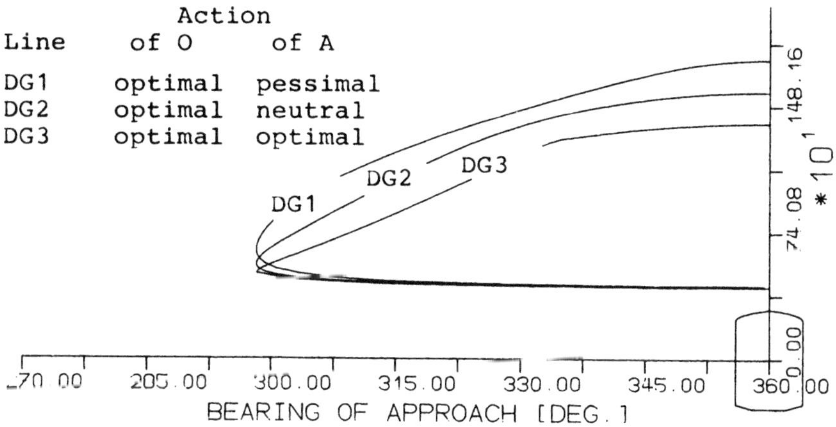

Fig. 8 Comparison of MMR and MR for different maneuver intention of A

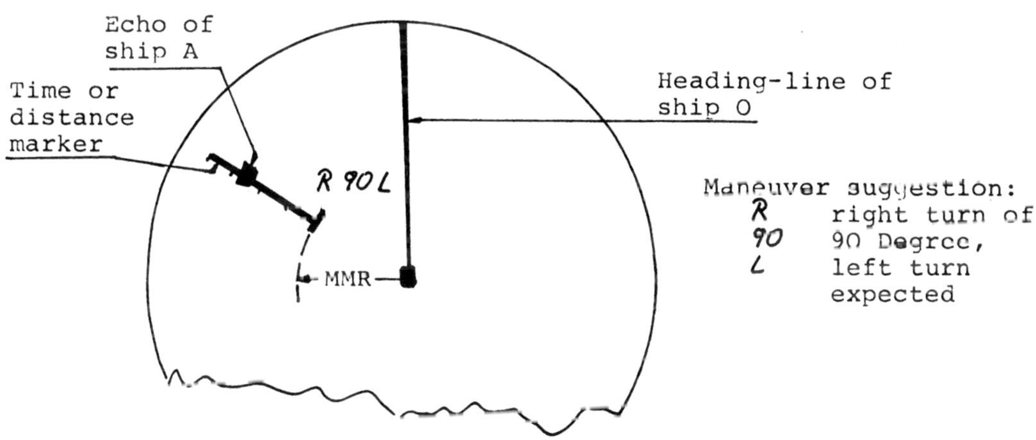

Fig. 9 Display of MMR and maneuver suggestion

The Measurement and Assessment of some Critical Skills of Mariners on CGI Shiphandling Simulators. Can the Mathematical Models meet the Demand?

P.M.P.MUIRHEAD

The Australian Maritime College
Launceston,Tasmania.
Australia.

Summary

The advent of the CGI shiphandling simulator has led to the development of extensive training programmes for Mariners in many aspects of seamanship and navigation. As a further development,this paper looks at the way in which the technology may be used to supplement existing methods of assessment of certain practical skills, and asks if the tests on which most models are validated, are sufficiently broad in scope to meet such a demand. The author has drawn upon some experiences gained from current programmes being researched at UWIST,Cardiff, and SUSAN, Hamburg.

ASSESSMENT AND LEVELS OF COMPETENCE.

Competence is a more commonly used word for competency.It can be described as a condition of being capable. Most dictionaries define the word to mean possessing ability and sufficient skill for the purpose in an accomplished and authoritative manner. Showing competence implies that a person displays a capability in one form or another to discharge a function effectively and satisfactorily.

Seamanship is a skill requiring a practised ability by a skilled person. To attain such a capability, a person must be highly trained or experienced or both. The definition is clear but what does it mean within the context of a competent seaman?

The role of the seafarer involves the use of many different skills. From the earliest days, the response of the seafarer to a series of oral or written questions has determined,in the eyes of the authorities, his competence or otherwise. The difficulty lies not in defining competence, but in discovering firstly which parameters will be used to set the level of performance expected, and secondly how to measure performance against them to adjudge that competence has been attained.

In the traditional British and Australian systems of oral examinations, what is it that the examiner seeks from a candidate? For deck officers, the British regulations specifically state that :

> *"The oral examination is intended to ascertain a candidate's competency in the practical aspects of an officer's duty"* [1]

The Principal Examiner of Masters and Mates (UK) has said that the need to satisfy examiners as to a seafarer's competence in matters laid down in the seamanship syllabus in particular remains of paramount importance. In this regard, the examination should relate to good seagoing practices and provide the means of ensuring that an adequate standard of competence in those charged with the navigation and safe management of the ship is maintained. [2]

The establishment of a seafarer's competence in the area of shiphandling for example is determined by an examinee moving a model of a ship on a flat surface and describing the helm, engine and ship movements likely to be taking place during the manoeuvre. A clear explanation may lead the examiner to conclude that the candidate understands the principles and could carry out the task in practice if required. Many other practical skills laid down in the syllabii for certificates or licences will be so judged. In the USA and Europe, the determination of such skill levels is arrived at by written assessment. Few candidates at watchkeeper or master certificate level have any practical shiphandling experience. It is very much a case therefore of relating a plausible chain of events without any substantiation to the claimed end results. The shiphandling simulator of itself may not necessarily transfer all end results to the real world, although early research indicates that such a transfer could take place to a high degree (Miller, Saxe, D'Amico 1982); (Meurn, Williams 1984). It does however provide a much needed opportunity for the seafarer to put theory into practice. If the level of ship model fidelity is high then the possibility of subjecting the seafarer to realistic critical task skills in many situations and conditions remains feasible. Importantly, the confidence and belief of the user in the validity of his actions and in the end results will increase. Without this, the seafarer will remain sceptical of its use both to acquire practical skills, and to make a judgement of his capabilities.

Lacking other means of testing, the traditional system has proved quite effective over the years, despite suffering the disadvantages of some lack of uniformity and standardisation as an assessment method. The great drawback to the oral or written assessment method lies in its inability to test, in a physical or hands-on sense, many of the practical skills in the seamanship area, upon which lie most heavily the onus of proof of competence in its fully defined sense. Introducing new technology in the form of shiphandling simulators has provided a possible supplementary weapon in the armoury of the examiner to counter such limitations.

THE ROLE OF SIMULATOR ASSESSMENT IN EXAMINATIONS

Neither written nor oral assessment methods give any assurance that shiphandling skills in particular can be carried out safely in practice, or that the opportunity to put them to use will not be many years hence. The advent of the shiphandling simulator has provided an avenue of riskless training in such skills for the seafarer, overcoming the possible long wait before experience is gained in the real situation and the consequent attendant risk to ship and life when reality comes upon them. A number of skills can only properly be viewed and assessed by means of a practical demonstration. By definition it includes those skills needed by the master or watchkeeper to manage and operate a ship safely and efficiently at all times. In particular it calls for the ability to handle and manoeuvre a ship, to a degree compatible to the grade of certificate sought. Can we continue to attempt to ascertain a candidate's grasp of fundamentals in these areas by oral and written assessment in the face of new technology? It cannot be ignored. At the same time it is difficult to put forward the argument that simulation can replace traditional assessment methods. The cost and time involved to train and assess a seafarer in all syllabus requirements is clearly beyond both existing technology and the budgets of training and simulation centres. The ship handling simulator does however provide a medium through which a candidate can demonstrate in a practical way whether he does or does not possess sufficient skill and ability to accomplish certain set tasks effectively and satisfactorily.

The idea is not new, having been used for many years in the aircraft and other industries. Despite the innate conservatism of the mariner, a form of assessment has been in force for some years in many countries by means of mandatory radar and navigational aids short courses. For example, many critical navigational skills are acquired by active participation in radar observer,simulator and ARPA courses, although formal assessment may not be part of the course. Their value is recognised however by the comment in the UK that *"The subject matter is considered to be vital to safety and a level of proficiency can be achieved without the examiners becoming involved except in a supervisory way to see that proper standards are maintained".* [2] With the shiphandling simulator, the examiner or appointed assessor has a unique opportunity to observe at first hand if a candidate possesses the basic skills and ability to perform essential seamanship tasks in an effective, safe and satisfactory manner.

SKILLS FOR ASSESSMENT

The type of critical skill that can be simulated, tested and evaluated on the shiphandling simulator is limited.Constraints of time,student numbers, the setting of task limitations and influencing variables, and lastly and most importantly, an ability to monitor the overall performance, place restrictions upon the number of practical skills that can be assessed.With this in mind an assessment programme has been established, the objective of which is *to establish whether a seafarer possesses a fundamental grasp and understanding of certain practical seamanship and navigational safety skills at his prescribed level of grading for a certificate or licence of competence as a watchkeeping officer or ship's master.*

The list of seamanship skills of relevance to this programme has been selected from the following syllabii or conventions, covering basic watchkeeper grades, and chief mate/master grade.

(a) IMO S.T.C.W. convention Regulation II/4 - Officers in charge of a navigation watch on ships >= 200 g.r.t.
(b) IMO S.T.C.W. convention regulation II/2 - masters and chief mates of vessels >= 200 g.r.t.
(c) Master class 1,2,3 (U.K.).
(d) A.G.W (Wachoffizer auf Grosser Fahrt)-West Germany.
(e) A.G.(Kapitän auf Grosser Fahrt)-West Germany.
(f) Master class I/II (Australia).
(g) Second mate class I/II (Australia).

WATCHKEEPER LEVEL

The tasks have been selected on the basis of forming core seamanship skills at the grade, of being practical skills that can more easily be evaluated on the simulator than through the medium of the oral/written examination, and for their reality in performance in the simulator medium.

1. Conn the ship safely, with the use of recognised helm orders.

2. Approach an anchorage and bring a ship to a single anchor.
3. Enter and make a passage in a buoyed channel.
4. Turn a vessel (single screw) short round off a berth (no wind or tide).
5. Turn a vessel (twin screw) in a channel (no wind or tide).
6. Berth or unberth a vessel(single or twin screw) without tug assistance, with or without bow thruster (no wind or tide).

CHIEF MATE/MASTER LEVEL

Again the tasks have been selected for being core seamanship skills at chief mate/master level, and which can be recreated to a high degree of fidelity on the simulator with exposure to environmental effects required at this standard.

1. Manoeuvring characteristics and capabilities of different ship types and their application to shiphandling operations.
2. Anchor or moor a vessel with wind and/or tide effects.
3. Pick up or drop a pilot with wind and/or tide effects.
4. Narrow channels: Pass close to other vessels (interaction)
Manoeuvre at deep draught (squat,roll,pitch effects)
Control a passage (bank effect, helm response, engine kick effect)
5. Berth and unberth a vessel (single or twin screw) with wind and tide effects.
6. Use tugs in shiphandling and berthing operations.
7. Handle a disabled vessel in adverse conditions.

SETTING THE TEST PARAMETERS.

Whilst the oral examination is totally subjective in its method of evaluation, the shiphandling simulator provides a more clearcut means of establishing standards of performance. Provided adequate monitoring facilities are available, considerable quantitative measurement of the exercise can take place. In addition the assessor has the advantage of his visual perception to assist in judgement making, as well as observing the perceptual-motor skills of the candidate at work.

The intention is not to assess performance solely by measuring ship movement within pre-conceived limits, or to state that the test action does not mirror methods laid down in good text books! Rather, efforts are being concentrated to perceive the way in which a task is performed in a practical, safe and effective manner. The test exercises have been run by mariners with differing backgrounds and levels of experiences, the results being recorded by plotter,data printout,tape recording and visual observation. This has been drawn upon to establish some broad parameters which the trainee can reasonably be expected to meet, taking into account the experience level at a particular grade. The examiner/assessor will make both a subjective and quantitative assessment of the performance based on the evidence provided by visual, aural and recorded observation, and the resulting final situation.

The effectiveness of a manoeuvre is not easily established, the results being affected by a number of variables. The design of the exercises has been such as to attempt to keep these disturbing influences to a measurable minimum. If the end result correlates with known experiences, particularly if the task is conducted safely and in a seamanlike manner, then there would be some justification for concluding that the performance had validity from a competence point of view (I refer you again to the definition at the beginning of this paper). If however a manoeuvre results in the vessel being placed in a position of risk, or takes an inordinate length of time to complete, then its effectiveness must be questioned. An attempt has therefore been made to include a parameter of time, within which candidates ought to complete the task or exercise, bearing in mind the level of experience, ship characteristics, location and influencing dynamic factors.

MATHEMATICAL MODELS FOR ASSESSMENT.

One of the most important features in any programme of this type, lies in the ability of the own ship model to mirror the real ship manoeuvres, when the seafarer places such demands upon it. The level of fidelity must be high and calls for a sophisticated model to meet the seamanship skills to be tested at the watchkeeper and chief mate/master levels. It must be capable of simulating all slow speed manoeuvres, shallow water effects and bank and passing effects. Most models employed on the world's shiphandling simulators are now provided with these effects. However in my experience having driven ship models on simulators in the USA, Britain, Germany, Australia and Norway, no one mathematical model seems to have achieved the ultimate in reality.The real problem in providing a sophisticated slow speed manoeuvring model for this type of work lies in the trial data that is available for the real ship. upon which the modeller designs and fine tunes his simulated version. Outside of the well known shallow water manoeuvring trials carried out on the VLCC's Esso Bernica(1973), Magdala, and Esso Osaka (1977) [3], little information is available on the behaviour of any type of vessel at slow speed in shallow waters (h/T=1.2).

Nearly all manoeuvring trials on new vessels are carried out in deep water and follow an established set of standard tests, which are linked to recent requirements for such information to be readily available onboard ship. These trials usually include turning tests to port and starboard at full speed and slow speed, stopping tests when going ahead, coasting stop test, Z-manoeuvre, Reverse spiral test, and bow thruster test (if fitted). [4], [5], [6]. Whilst data in full load and ballast conditions is usually provided, it may not always be comprehensive.

Yet the function of the full mission shiphandling simulator is to mirror the actions of a vessel within limited waters, where movements generally must be slow and controlled. Where are the test trials that will provide information on a ships handling characteristics in these circumstances? They are not available! The modeller must develop by trial and error and by subjective evaluation through the eyes, senses and feel of experienced mariners, an own ship model to perform an increasing complexity of shiphandling tasks. As the number of shiphandling simulators in the world continues to grow (26 by the end of 1986), so will the demand grow for them to meet more exacting training and developmental tasks. There are limitations on the reliability of this method.

To meet the assessment programme in full, what must the model be capable of, and can we validate its performance? Some of the tasks set in the programme call for the seafarer to perform the following:-

1. Turn a 280 metre single screw ship short round in an swinging area of 600 metres in diameter; berth the vessel port side to a jetty (no wind or tide effects).

2. Turn a 214 metre twin screw vessel round in a channel 400 metres in diameter; bring accurately to anchor in a designated position.
3. Take a 180 metre single screw CPP vessel off a berth, swing, and proceed down channel, meeting an inbound deep draught vessel on the way out.
4. Turn a 290 metre VLCC off a jetty with up to 4 tugs in an area 600 metres diameter with strong offshore wind and tidal effects.

Very little of the manoeuvring data of the ship is of use in these situations. The provision of low speed equations on the models is based on well founded mathematical principles, and with adjustment can be made to perform like a real ship. But is it the real ship in this type of manoeuvre? Some low speed models still exhibit unreal tranverse thrust effects when put through their paces. A solution would be to establish a range of manoeuvring tests for vessels at slow speed in restricted waterways. Such tests should relate to the type of operations likely to be carried out by the type and size of vessel. All simulator own ship models should be provided with anchors, not only to hold the ground, but which can also be dredged across the ground under tension for use in many berthing manoeuvres. Much development remains to be carried out to improve the simulation of tug forces. The control of tug heading, towing direction and tension,and attachment point on the ship still lacks realism, for both operator and instructor, particularly when handling four or more tugs. Even the excellent Krupp -Atlas system (figure 1) requires considerable dexterity on the part of the operator.

When carrying out berthing manoeuvres, the visual aspects of the jetty on the simulator assume even greater significance. Field of view of at least 100 degrees on the berthing side is,in the authors experience, the minimum visual aspect needed, and performance can be improved by providing a stern view monitor on the bridge aft. Field of view can be increased by off-setting the forebody to one side, but this imparts too great a sense of unreality into the exercise. Contrast levels between water, jetty, land background, and ships hull have also proved to have a strong influence on student performance. Colour selection needs careful thought! The problem of getting alongside is currently solved by accepting a result of ship stopped and parallel to the jetty within 50 metres (inexperienced seafarers) and 25 metres (at the more experienced levels). Figure 2 illustrates the difficulty. The addition of a cushioning effect at the jetty will solve this problem, and Hamburg has modelled a jetty that provides ship rejection above a certain rate of approach.

FURTHER TESTS FOR MODEL AND EXERCISE FIDELITY

Experience with the testing programme currently being carried out at CASSIM suggests that the establishment of performance parameters for shiphandling manoeuvres at slow speed could be greatly improved if some data of actual manoeuvres on ships within the size of vessels modelled could be made available. Observation of student performance and post assessment comments indicate that even the most inexperienced student could achieve better practical results if the characteristics of the vessel at low speed were more readily available beforehand.The opportunity to extend the known characteristics of a ship to include the slow speed manoeuvring regime should be grasped by all seafarers whenever possible.To this end, a number of basic slow speed approach and turning manoeuvres are suggested. Data in the ballast condition is often lacking for manoeuvring operations.

AUSTRALIAN MARITIME COLLEGE

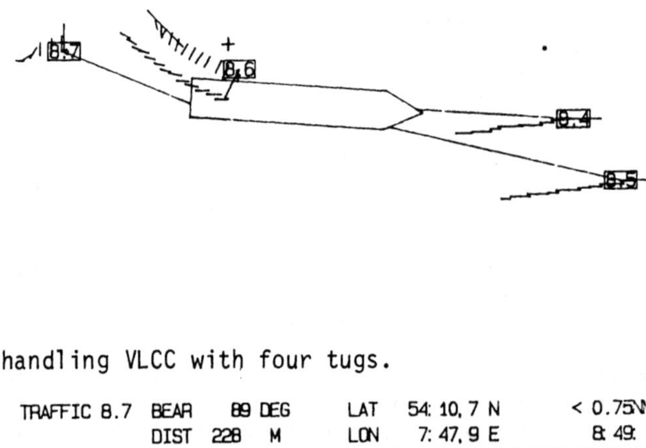

figure 1. handling VLCC with four tugs.

figure 2..berthing starboard side to at SIMLAND..

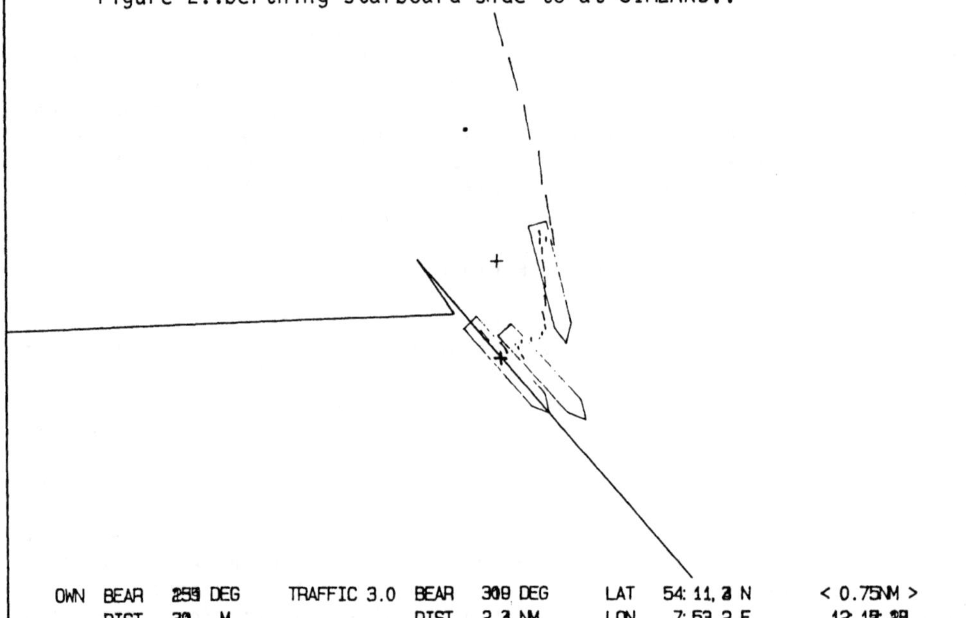

(a) TURN SHORT ROUND (SINGLE SCREW)
In a turning area limited to a diameter 2.5 to 3 times the length of the vessel,turn a single right hand propeller vessel through 180 degrees using helm and engine only, from a line of approach to the left side of the area with speed of about 4 knots.Conditions should be slack water and little or no wind force (maximum 2). Reverse the direction of turn for left handed screws. For vessels with single CPP screw the tranverse effect will be felt in the opposite direction.

For vessels fitted with a bow thruster the above tests should then be repeated using the thruster at maximum power. The diameter of the turning area used, the rates of turn achieved and the time taken should be noted, and all helm,engine and thruster movements recorded, with related RPM, pitch, speed and headings recorded in each case.

(b) RESTRICTED TURN (TWIN SCREW)
In a turning area limited to a diameter 2 times the length of the vessel, turn twin screw vessels through 180 degrees first to port and then to starboard using helm and engines only, using maximum power in the first trial, and a maximum of slow ahead, slow astern in a second trial. The line of approach should be with the vessel stopped in the centre of the turning area. Slack water and little wind force should prevail. If bow thruster is fitted, repeat the foregoing tests with thruster at full power.

The movement of the vessel within the area should be measured and all helm and engine orders recorded. The rate of turn, RPM, speed, pitch and heading should be recorded with these movements.

(c) SLOW SPEED STOPPING/PULLING UP

From a straight line of approach at speeds in the range of 2 to 6 knots and engine order telegraph on stop, the headreach, lateral deviation, final heading and rate of turn (if any) should be recorded at slow, half and full speed astern. The time taken from the first astern order to the moment when the ship is stopped dead in the water is to be noted. With twin screw vessels, both engines are put astern together. Many vessels hold manoeuvring data from which this information can be extracted in the deep water situation. At SUSAN - Hamburg, Herman Von Morgenstern has carried out tests on the single screw container vessel in shallow water conditions, and an example of a test of the type suggested is shown in figure 3. What is lacking of course is the real ship data in shallow water to make a comparison and an evaluation on the validity of model behaviour in such conditions.

(d) SLOW SPEED TURNS AT 10/20 DEGREES OF HELM.

Turning circle data provided onboard relates to full speed, full helm over conditions in most instances. Longman and Douglas [7] have clearly shown that the seafarer uses a much greater variety of rudder angle to take avoiding action at sea, and have suggested additional tests and data for these conditions.
For shiphandling in restricted waters, the following additional tests are suggested. With the vessel at dead slow ahead apply 10 degrees helm and note the advance, transfer and time for 90 degrees change in heading. Repeat with 20 degrees helm. Record the effect of such helm movements for

figure 3.

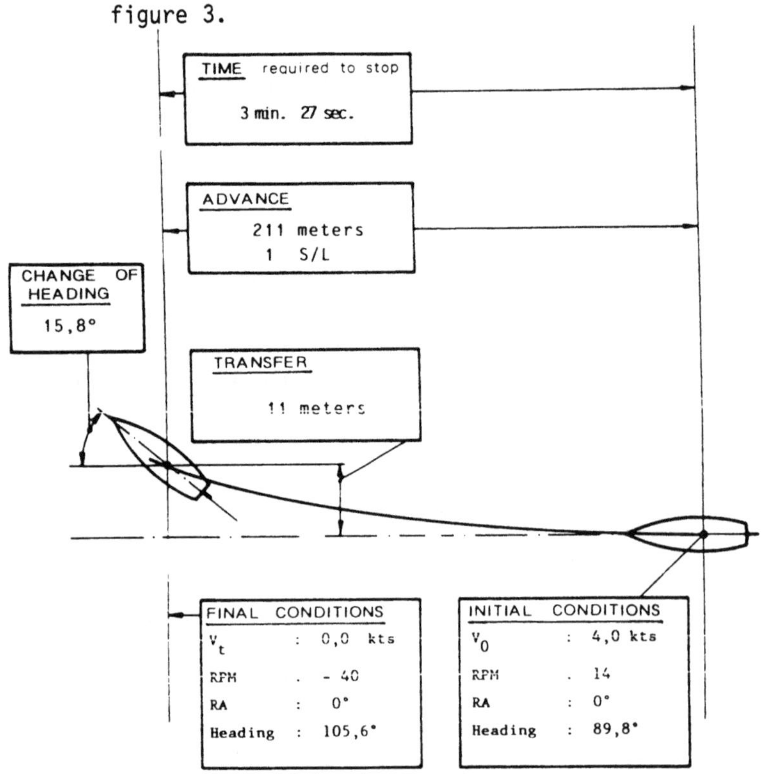

Reprinted by permission of Fachhochschule,Hamburg.

FACHHOCHSCHULE HAMBURG
FACHBEREICH SEEFAHRT

Time	Order	Remarks
0'00"	Sl-Ast.	
0'02"		Eng. rev.

Load.Cond.: Full   Shallow Water

MANOEUVRING DATAS   CONTAINER - VESSEL

MANOEUVRE:

4 kts. AHEAD - SLOW ASTERN TO DEADSTOP

ENVIRONMENT and STARTING CONDITIONS :
Wind: —   Current: —
Sea : calm   Visibil.good

Page: 12
Of:

slow and half ahead. The rate of turn, and speed should be recorded at appropriate intervals (20 degrees).

In all the above tests, it is recognised that the range of tests carried out will be determined to some extent by the power and size characteristics of the ship. Most are within the bounds of practicality for the shipmaster.

The provision of much of the above data would allow greater fidelity in model development and a more realistic result in many of the common berthing manoeuvres carried out in practice. The behaviour of individual twin screw vessels is generally poorly documented at slow speeds and the provision of extra data would help to overcome this. Whilst recognising that time and cost do not always permit the inclusion of additional tests at new ship trials, many of these tests could be carried out during the course of a vessel's commercial life. Those institutions fortunate enough to operate training vessels in their own right, have the opportunity to carry out some pioneering work in this field. The Australian Maritime College will be able to contribute to modelling fidelity by comparing new test data from its real ship with the mathematical model results on the shiphandling simulator. A twin screw 64 metre length vessel is a small start but the results will be invaluable. Arrangements with shipowners for the additional testing of simulator ship model counterparts could greatly assist the modeller, the seafarer and the regulatory authority to establish acceptable levels of shiphandling performance. If the simulator is to be an accepted training medium, it must satisfy all sections of the industry as to its validity.

REFERENCES

1. HMSO : Regulations for the examination of Masters and Mates.1982.

2. Jestico A.M. Developments in the examination of Masters and Mates. Seaways, Journal of Nautical Institute, 6/1984.

3. Crane C.L Manoeuvring trials of 278,000 DWT tanker in shallow and deep waters. S.N.A.M.E. 1979.

4. Det Norske Veritas. Navigational Aids and bridge systems: manoeuvrability tests. 1974.

5. Patterson D.R. On the steering and manoeuvring and stopping of ships. International Marine Safety Symposium 1980.

6. IMO. Manoeuvring characteristics of ships: Guidelines for manoeuvring performance in ship design. 1984.

7. Longman.M. Douglas J.D. Presentation of manoeuvring data. Ship Operations and Safety Conference 1981.

Session 11 Under-Water Vehicles Operation

A Simulator for Remotely Operated Underwater Vehicles

G.K. KAPSENBERG

TNO-IWECO
Delft, The Netherlands

Summary

This paper outlines the future need of simulation techniques for Remotely Operated underwater Vehicles (ROVs).
Simulation techniques are here applied to predict the manoeuvring behaviour of a ROV. The derivation of the mathematical models is described. It is further shown how these mathematical models are implemented on a microcomputer, and how these are integrated for application in a basic simulator.

Introduction

In the second half of the 1970's, a rapidly increasing number of Remotely Operated Underwater Vehicles (ROVs) have been used by the offshore industry in the North Sea. At first, these vehicles had to prove their worth (especially in an economical sense) in competition with saturation divers, but they rapidly proved to be competitive for a limited number of tasks. These tasks mostly contained inspection work like the reconnaissance of a pipeline trajectory, visual inspection of a pipeline or a jacket and assistance of saturation divers (visual feedback to the surface, providing light).
In this first period, the market was relatively easy. The challenge was accepted by a large number of companies, who each developed their own type of ROV. Only few of these were so successful that they saw their series number increase above 10.
Now, halfway through the 1980's, the ROV market is almost saturated. Every new ROV system must promise a technical advancement in comparison with its predecessors and its competitors. This advancement is mostly achieved by concentrating the design development on a specific task; the era of the dedicated ROV has started.
In the coming period, the dedicated ROV has to prove itself as a reliable replacement of the saturation diver. The saturation diver has played a

vital role in the accomplishments of the offshore industry, but because the work theatre in the North Sea is moving north to deeper and more hostile waters, the work of the saturation diver is nearing the end of the capabilities of a human being.
The dedicated ROV has to take over all the jobs which are now executed by divers; this not only includes inspection work, but also maintenance of subsea constructions. There is no doubt that the maintenance type of ROV will be fitted with one or more manipulators, which will consequently increase the workload on the ROV-pilot. This further implies that the ROV pilot need to have a training period to fulfil his tasks successfully. As a proof of the workcapability of a pilot, there comes a time that oil companies and/or classification societies will demand a formal training period and a licensed pilot.

In conclusion, we predict:
- Future ROVs will be designed with a specific task in mind;
- Specific ROVs will need a specially trained pilot to live up to the expectations;
- There will be a need for internationally recognized training facilities for basic pilot training.

This last need has been recognized by Det Norske Veritas [2,3], and they play an active role in the development and operation of the NUTEC-bay training facility. This training facility gives the opportunity of acquiring:
- technical/theoretical knowledge
- basic flying techniques
- underwater inspection techniques.

The training area in the bay is equipped with several parts of offshore underwater constructions; the trainee flies an actual ROV. In our opinion, such a training facility could be well supplemented by training on simulators. Simulators have advantages with respect to training on the real thing. These advantages are:
- repeatability of a training excercise under exactly the same circumstances;
- easy introduction of malfunctions by the instructor;
- training for emergency situations
- possibly costs.

The heart of every simulator is the mathematical model of the simulated object. The mathematical model can also be used for other purposes as, for instance, evaluation of a design. In this respect, it is possible to execute a standard series of manoeuvres/tasks and to evaluate the results. We are convinced that this last application of simulation techniques is also a valuable contribution towards the development of the hi-tech ROV.

As a start in the development of a versatile simulation package, we have derived a mathematical model of one specific ROV and we have built a simulator around it for demonstration purposes.
This paper deals with the mathematical model of the ROV, its implementation on a microcomputer and the additional components to make it a simulator.

The simulated ROV

The simulated ROV is a small vehicle, designed, built and operated by the Dutch Heerema Group. The ROV is called the SUB 300, and it is used for pipeline inspections and comparable tasks.
Its most striking feature with respect to other ROVs is its suitability for operations at high current velocities. The thruster-power/weight ratio is very favourable, and also most important, the thrust is high in every horizontal direction.

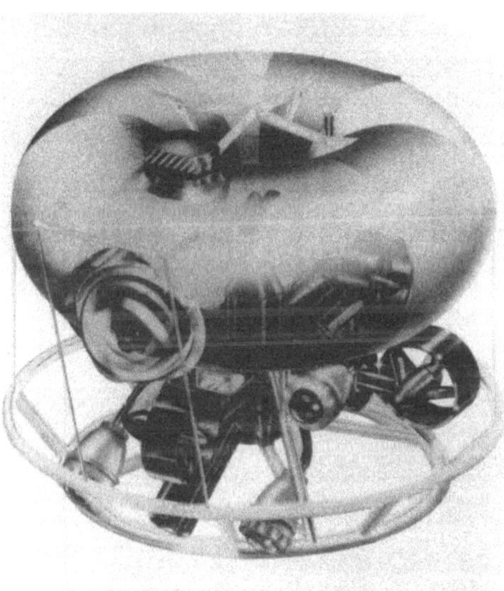

Manufacturer	: Heerema Innovation Eng. Leiden, The Netherlands
Build year	: 1984
Dimensions	: Ø 1.25m x 1.00m
Weight dry	: 500 kgf
Payload	: 80 kgf
Structure	: open frame with cast aluminium doughnut shaped pressure housing on top
Thrusters	: 4 rpm-controlled thrusters in hor. X-Y configuration, 1 vert. thruster in doughnut center
Power	: thrusterpower 5x5.5 kW

Figure 1. Artist impression and general data of SUB 300.

Another basic feature of this ROV is that it is operated from a garage. Using this system, the connection to the surface is not limiting the freedom of the ROV in the same way as it did with the older system of one continuous umbilical to the surface support vessel.
The ROV is equipped with video-cameras (colour, black & white), sonar, lights and an automatic heading and depth control system.

The mathematical models

We have tried to construct the mathematical models of all ROV components in as modular way as possible. This modularity has the advantage of easy adaption to different ROV geometries; it makes also the replacement of modules by improved versions a straight forward operation.
The modules of the ROV simulation program are shown in figure 2. Most of them calculate external forces on the ROV system components. All forces on the ROV are substituted in the six degrees of freedom equations of motions. An integration routine completes the simulation loop.
The contents of the modules is now briefly described:

- module 1

  Contains the state vectors $\underline{x}$, $\underline{\dot{x}}$ at time level k.

  The state vector $\underline{x}$ is defined: $\underline{x}^T = (x, y, z, \phi, \Theta, \psi)$

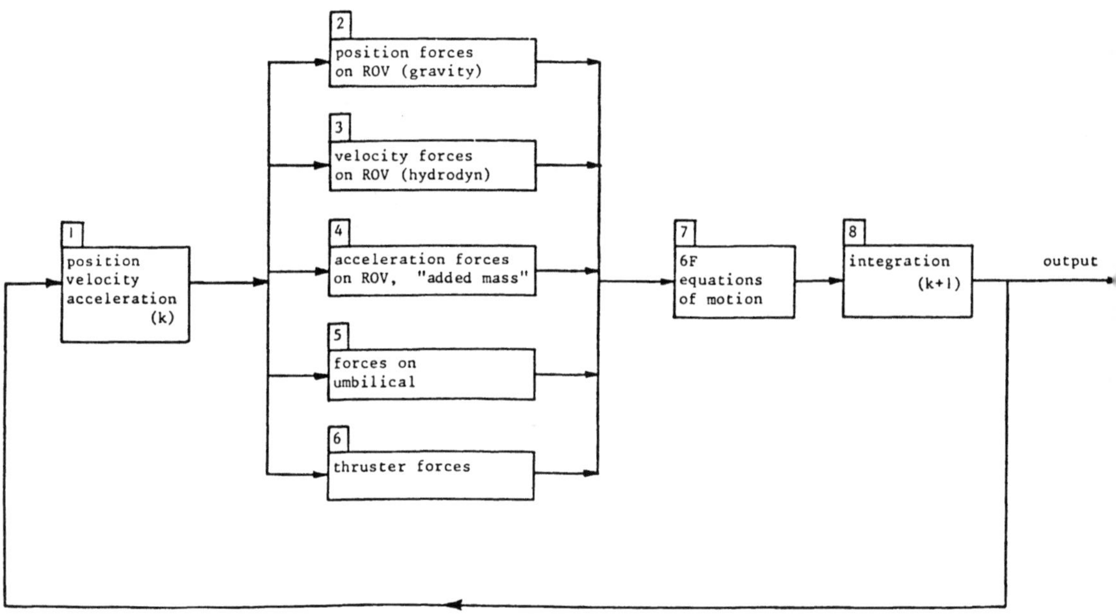

Figure 2. Flow diagram of simulation program.

- module 2

Contains the static forces on the ROV. This ROV has positive buoyancy for reasons of safety at a "black-out". This positive buoyancy force (FG-FB) is transformed to the ROV system of axes by multiplication with the rotation matrix R. The rotation matrix R is defined to transform vectors from the earth-fixed system of axes $X_oY_oZ_o$ to the body fixed system XYZ.

- module 3

The velocity forces are constructed from lift and drag data of basic elements. These lift and drag forces are based on data in Hoerners work [5,6]. The forces on the basic elements of the ROV are added to the estimated interaction forces.
In this way, the hydrodynamic forces are calculated for a limited number of cases (α,β); intermediate values are interpolated.
In formula:

$$\underline{F}_{vel} = \tfrac{1}{2}\rho\, U^2 \sum A\, C_D\,(\alpha,\beta) + \tfrac{1}{2}\rho\, U^2 \sum A\, C_L\,(\alpha,\beta) + \tfrac{1}{2}\rho\, \omega\,|\omega|\, C_\omega$$

in which:

ω - angular velocity of vehicle

coefficients $\sum A\, C_D\,(\alpha,\beta)$ and $\sum A\, C_L\,(\alpha,\beta)$ contain the sum of the products of relevant area and drag/lift coefficient of each basic element including interactions. The coefficients can be found in look-up tables.

- module 4

Apart from the virtual increase of the mass of a submerged vehicle, the "added mass" phenomenon has more effects. These effects can be determined if the equations of motion are derived for a moving system of axis with the "total mass" matrix (total mass = mass + added mass). This derivation was, for instance, published by F.H. Imlay, [7]. For the X-equation, this results in:

$$\begin{aligned} FX_{acc} = {} & X_{\dot u}\dot u + X_{\dot w}(\dot w + uq) + X_{\dot q}\dot q + Z_{\dot w}wq + Z_{\dot q}q^2 + \\ & + X_{\dot v}\dot v + X_{\dot p}\dot p + X_{\dot r}\dot r - Y_{\dot p}rp - Y_{\dot p}rp - Y_{\dot r}r^2 + \\ & - X_{\dot v}ur - Y_{\dot w}wr + Y_{\dot w}vq + Z_{\dot p}pq - (Y_{\dot q} - Z_{\dot r})qr \end{aligned}$$

The symmetry of the added mass matrix has been used in this expression. Any symmetry of the considered vehicle is not included. Even if the vehicle has only one plane of symmetry, this expression is significantly reduced. For this ROV, which is nearly axi-symmetric with respect to the Z-axis, we used:

$$FX_{acc} = X_{\dot{u}}\dot{u} - Y_{\dot{v}}\, vr + Z_{\dot{w}}\, wq$$

- module 5

For the calculation of the forces acting on the umbilical, and most important for the forces of the umbilical on the ROV, we have developed two models.
Model 1 uses a "finite segment" approach: the umbilical is divided into a number of segments. Each of these segments may be curved, but has its mass concentrated in the center. This point is also the center of all external forces. This model gives a fairly accurate description of the whereabouts of the umbilical and of the forces exerted by the umbilical on the ROV. It is, however, not very suitable for real-time simulation on the applied micro computer.
For our simulator, we needed a more simplified model of the umbilical. Since our ROV uses a cage-tether system it seems to be acceptable to use a quasi-static approach.
With this approach, the tether is assumed to have a parabolic shape. The parabola is defined by the locations of ROV and garage and the length of the tether. The plane of the parabola is defined by the relative velocity vector of the tether through the fluid and the line which connects the ROV and the garage.
The dynamic simulation of the tether motions is thought to consist of a series of static situations.

- module 6

The thruster routine is one of the more complex routines. The complexity is not brought on by the fact that there are five thrusters, it is a consequence of the mutual interaction of the thrusters. The four thrusters for the horizontal plane motions are situated close together, and the outflowing fluid of one thruster has a big influence on the neighbour thruster.
This interaction process is solved in the mathematical model by a two

step iteration process. The thrusters are first calculated without any interaction, this calculation determines outflow velocities of the various thrusters. The second step takes the outflow velocities of the thrusters into account to determine the magnitude and direction of the incoming flow for each individual thruster.
This second step determines the final thrust, torque and sideforce production of the thrusters.
The propeller diagram used for this module is a so-called four quadrant diagram [12]. Such a diagram gives thrust and torque coefficients as a function for the hydrodynamic pitch angle β. The influence of the angle of incidence of the incoming flow is taken into account by experimentally determined multipliers [11].

The vertical thruster is a special case. This thruster is situated at the center of the buoyancy volume, and cannot be considered as a thruster in a nozzle at an angle of attack. The sideforce production calculation of the four horizontal thrusters is changed for a "momentum drag" calculation.
If the ROV is flying forward, and the vertical thruster is operated, the flow is forced from the horizontal plane through the vertical thruster. The reaction of this force on the ROV is called momentum drag. A formula for the momentum drag of the described case is:

$$F = \frac{\partial m}{\partial t} \cdot v = \rho\, A\, v_{pd}^{2}$$

in which:

v_{pd} - flow velocity through propeller disk.

The thrust of this vertical propellor is calculated in the same way as is described for the horizontal thrusters.

- module 7

This module solves the equations of motion; the output contains the accelerations $\underline{\ddot{x}}$ at time level k.

- module 8

Module 8 contains the explicit integration formulas for calculation of $\underline{x}$ and $\underline{\dot{x}}$ at time level k + 1.

The simulator program

The simulation program as given in fig. 2 is extended with several additional routines to make it suitable for a simulator. These additional routines are:

- analog/digital input
- analog output
- thruster allocation logic (to translate the joystick position and signals from the heading control to commanded thruster rpm's)

and for our special case:

- image generation
- heading control.

The analog and digital conversion routines are machine-dependent and are not described in this publication.
The thruster allocation logic is straight forward and also not described; some more attention is given to the last two routines:

- image generating routine

  Due to the limitations of our graphics processor (it takes 800 ns to put 1 pixel on the monitor), we designed a simple computer generated image. The sea bottom is represented as a perfect horizontal plane with a 1 x 1 m^2 grid on it. This grid gives the pilot a feeling for horizontal speed and changes in course.
  Changes in depth are indicated by an increase/decrease in size of the grid on the monitor and a change in the perspective of the view.
  The range of visibility is adjustable, we mostly use a range of 4.0 m.

  Our simulator hydrospace is, apart from the bottom, limited by one vertical subsea wall. This wall enables the pilot to practice vertical flight.
  The simulation program has to run real time for the simulator. In order to present a smooth picture on the monitor, the update frequency has to be rather high. We achieved 8 Hz on our present microcomputer. This up-date frequency gives an almost smooth impression of motions on the monitor.

- heading control

  The simulated ROV is rather high powered and nearly axi-symmetric with

respect to the Z-axis. The combination of these qualities makes it very difficult to keep the ROV manually on course. Therefore, a heading control system is incorporated in the simulator.
The control system is of the PID-type; it makes use of a difference in course, a course rate of turn and an integrated course difference to determine the commanded moment. The thruster allocation logic determines thruster rpm's from this commanded moment.
The control algorithm reads:

$$FN_{com} = K_p(\psi_{setp}-\psi) + \sum_{n=1}^{k} \frac{K_i \Delta t}{2} \quad (\psi_{setp}-\psi)_n + (\psi_{setp}-\psi)_{n-1} \quad + K_d\dot{\psi}$$

This control algorithm is supported by several routines to guarantee a smooth initialization (from manual to automatic mode) and a controlled rate of turn if a new setpoint is entered.
The selected setpoint can be adjusted with ± 30 degrees by the third axis, rotation, of the joystick.

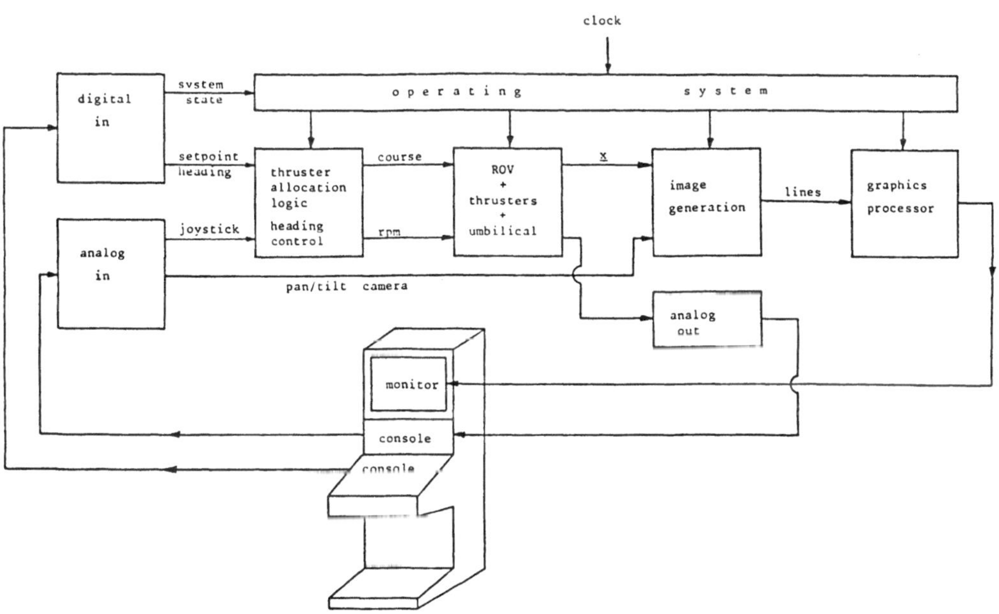

Figure 3. Flow diagram of simulator program.

The microcomputer system

The microcomputer used for the simulator is the versatile Geminix system, developed by a colleague TNO institute, TNO-IBBC. The system uses the Motorola MC68000 microprocessor both as master and slave processors on a standard VMEbus.

Additional hardware consists of memory, A/D and D/A converters, parallel and serial I/O, and a graphic display controller from NEC, the µPD7220. A hardware floating point module may be added to increase performance.

To reflect the modularity of the simulator system setup in the hardware configuration, a separate processor was used for each software module, see fig. 3. This results in rather limited dataflow between the four processes and allows easy interchangeability of software modules. By replacing only one software module, another ROV may be simulated, but also different consoles and/or video monitors may be connected to the simulator in the same fashion. It is also possible to 'isolate' a module and to develop and test it separately.

All software is written in the programming language 'C' and is, apart from some minor I/O drivers, transportable to another computer system.

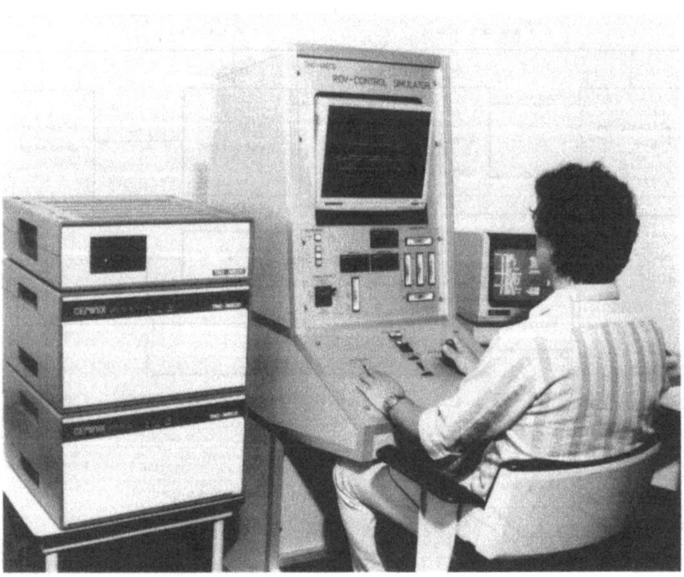

Figure 4. Operator console of the simulator (at a development stage).

The console

The console of the simulator, fig. 4, is a close replica of the one used for the real SUB 300. It is equipped with:

- 3-axes self centering joystick for X-Y motions and yaw
- thumbwheel for the Z-motion
- 3 thumb wheels for trimfunctions of horizontal plane motions
- setpoint selector for automatic heading control system
- 2-axes joystick for camera pan/tilt
- displays of
  - 5 thruster rpm's (analog)
  - heading setpoint (digital)
  - actual heading (digital)
  - depth (digital)
  - depth rate (analog)
- simulator mode selector (ON/OFF, INITIAL CONDITION, HOLD)
- monitor for the computer generated imagery.

Performance of simulator versus performance of ROV

Up to this moment there was no opportunity to conduct full scale measurements. The only comparison of the relative performance of the simulator was a very qualitive assessment by two ROV pilots with one year's experience with the SUB 300.

Their opinion was rather favourable, the "feeling" of the SUB 300 was recognized. Criticized was the rather poor information on the monitor of the distance to the bottom, and they judged the damping of the real SUB 300 in pitch and roll to be higher than that in the simulation. They concluded that a simulator could be very useful, even in this basic configuration, as an aid for basic pilot training.

Future developments

Looking at the future, we foresee separate developments for the different types of ROVs.

1. Light inspection work (pipeline trajectory reconnaisance, check on the burying process of pipes).

This vehicle will be relatively small and high powered, capable of speeds in the region of 5 knots. The visual inspection will then be replaced by some kind of acoustic sensoring. The acoustic sensoring of cracks (possibly concealed by marine growth) in concrete constructions is

already in an advanced state; it is the subject of a colleague TNO institute: TNO-TPD [14,15]. It is however necessary for these advanced types of sensors to move closely and at an exact distance above the inspected object. This can be achieved, for instance, by fitting the sensor at the end of an automated robot arm on the ROV. This automated arm has to compensate for small vertical deviations of the vehicle. Of course, mathematical modelling of vehicle dynamics and simulation techniques play an important role in the development of such a system.
Further developments could lead to more expensive autonomous vehicles.
These developments are presently studied for application in Arctic conditions, it seems to us to be a long way before they are applied in the North Sea.

2. 'Heavy' inspection work (inspection of jackets).
There is no doubt that the serious vehicle of this type needs to attach itself to the construction during the inspection. There are of course problems with the inspection itself, but recent advances on the cleaning of the relevant area, combined with worktools, sensors and cameras on a manipulator arm seem to solve most problems.
This leaves us still with the navigation problem inside the jacket until the docking at the desired location. The navigation can be assisted by feeding the ROV commands also to a simulation program which calculates a computer aided image as is described in this paper. This image gives the operator an artificial view of the construction, offering the advantage of infinite visibility.
It should also be possible to present the view of an imagined observer outside the jacket on the monitor. This view then shows the jacket, the vehicle and the umbilical. This could assist the operator if he has problems in finding the right way out of the structure.
More in the future we foresee a semi-autonomous vehicle without an umbilical. If such a vehicle is at its worksite, it could plug-in at its location on the jacket and make the connection to the control panel giving the operator the possibility to execute the inspection.

3. Work vehicle (maintenance of subsea constructions, replacement of components).
This type of vehicle should be versatile, so it is likely to be rather complicated. It is equipped with several manipulators and is unlikely to

be able to 'dock' at every worksite; sometimes only a support point must do.

It will be a complicated vehicle to operate effectively, so highly trained and qualified operators will be needed, making simulators a valuable tool.

This type of vehicle is already developed for the Shell Underwater Manifold Center; this UMC is a testbase in preparation of the move North to deeper waters in a hostile environment.

References

1. Brink, A.W., Spuyman, W. - TNO-IWECO: 'Remotely Controlled Underwater Vehicles'. Proposal for the improvement of RCUV design and operation techniques. TNO-IWECO report 5114007-82-1, October 1982.
2. Jensen, B., Svenning, B., Steiner, G., Evensen, C. - Det Norske Veritas: 'ROV pilot training'. SUBTECH '83, London.
3. Evensen, G. - Det Norske Veritas: 'ROV training is imperative'. ROV '83, San Diego.
4. Partridge, D.W. - Department of Energy, U.K.: 'Future Developments of ROVs'. Underwater Technology, Vol.10, no.1, 1984.
5. Hoerner, S.F.: 'Fluid dynamic drag'. Published by the author, 1965.
6. Hoerner, S.F.: 'Fluid dynamic lift'. Published by L.A. Hoerner, 1975.
7. Imlay, F.H. - David Taylor Model Basin: 'The complete expressions for added mass'. DTMB report 1528, July 1961.
8. Lewis, D.J., Lipscombe, J.M., Thomasson, P.G. - Cranfield Institute of Technology: 'The simulation of Remotely Operated Underwater Vehicles'. ROV '84, San Diego.
9. Miller, M.J. - Santa Fe Underwater Services Inc.: 'ROV testing & evaluation facilities'. ROV '84, San Diego.
10. Tolman, F. - TNO-IBBC: 'The Geminix workstation'. CAPE '83 Amsterdam.
11. Beek, J. v.d., Amersfoort, H.C. van - Netherlands Ship Model Basin: 'Experimental investigation into the hydrodynamic characteristics of a 1975 HP thruster unit'. NSMB report 0968-1-DT, November 1975.
12. Lammeren, W.P.A. van, Manen, J.D. van, Oosterveld, M.W.C. - Netherlands Ship Model Basin: 'The Wageningen B-screw series'. Schip en Werf, Vol.37, no.5, 1970.
13. Thomasson, P.G. - Cranfield Institute of Technology: 'Simulators for use as design aids and for operators training'. SUBTECH '83, London, Paper 9.1.
14. Verhulst, K. - TNO- TPD: 'Acoustical inspection of offshore constructions (in Dutch)'. Cement 1984, no. 3, page 173-176.
15. Schmidt, D.P., Vogel, J.A., Verhulst, K. - TNO-TPD: 'Development of ROV operated sonars for very high resolution imaging during underwater inspection'. ROV '84, San Diego.
16. Kapsenberg, G.K. - TNO-IWECO: A step towards the introduction of simulation techniques in the world of remotely operated underwater vehicles. International Shipbuilding Progress, April 1985.

The Program DYSMAS/ELC and its Application on Underwater Shock Loading of Vessels

by W. Bergerhoff, W. Mohr, W. Pfrang, F. Scharpf

Industrieanlagen-Betriebsgesellschaft,
Ottobrunn, West Germany

Abstract

The simulation of highly dynamic processes concerning structural response, fluid dynamics and the behavior of coupled systems can advantageously be managed by an explicit finite formulation, which is able to handle the various nonlinearities appearing in the wide field of short-time physics. The program family DYSMAS has been developed step by step in connection with a great number of projects mostly in the field of defense technology. The DYSMAS code consists of the two stand alone processors DYSMAS/L, a FE-Lagrangian code, and DYSMAS/E, a FD-Eulerian code and additionally of DYSMAS/ELC in which both are coupled together in 2D and 3D.

The code takes into account large displacements and distortions by the formulation of the basic equations. Its extensive material description allows the treatment of large plastic flow including work hardening and strain rate dependency, as well as material failure and post-failure-behavior using progressive methods such as crack-opening or erosion models. The 3D multibody contact processor enables the DYSMAS-user to simulate impact, penetration and crash problems. Especially fluid-structure-interaction phenomena are distinguished by nonlinearities present both in structure and in fluid. The low compressibility of water in compression and its cavitation properties under tension have great influence on the transient loading of vessels subjected to underwater detonations.

In the recent years DYSMAS/ELC was applied to the simulation of a lot of underwater detonations against various ships and submerged structures. As an example this report covers the evaluation of a torpedo shot in the nearfield of a frigate-sized ship. The investigation embraces the detonation shock wave and its propagation in the fluid and in the structure as well as the mechanical damage to the ship hull, caused by the shock wave and by the dynamic pressure due to the gas-bubble-expansion. The results comprise the assessment of damage on equipment based on the evaluation of the shock-spectra, which enable the analyst to make a decision about the residual serviceability of the total system or subsytems.

DYSMAS/ELC at present is also used to simulate soil-structure interaction phenomena as for instance the detonation of buried charges against a shelter structure or terminal ballistics phenomena including ricochetting of projectiles on soil or fluid.

Introduction

Mathematical mapping of physical processes is a commonly used method in engineering to simulate real world sequences. The objectives of this effort are better interpretation of experimental results, illumination of complex physical correlations, determination of system parameters by estimation and extrapolation into fields, where experiments are impossible.

The entire simulation of an underwater detonation against the complex structure of a naval vehicle became pracţticable in the recent years due to the immense progress both in hardware and software development. Calculations of this type are mostly performed in defense technology to design shock resistent structures and equipment on one hand and to study weapon effectiveness on the other hand.

The Problem

Basically there are two different possibilities to gain realistic load assumptions which serve as input data for design or for further detailed investigations. First of all full-scale tests can be performed with well placed measuring points to obtain the arriving pressure transients at the ship hull as well as acceleration transients as base excitations of the equipment. These tests are not easy to carry out, at least they are very expensive. One can take remedial measures by simulating such physical phenomena numerically. Depending on the refinement of the discretisation used for the problem setup load data are available at nearly any point of the mapped region.

The entire computational treatment of underwater explosions against structures embraces three main components, namely

- the fluid dynamics part concerning detonation, shock wave propagation and gasbubble behaviour
- the structural dynamics part including the shock response and severe deformations
- the coupling part, which realizes the important fluid-structure interaction.

In principle a total uncoupled handling of such problems is possible and often used. The first step is to determine the load transients along a rigid boundary and in the second step these loads are applied to the structure. In this case no interaction effects are taken into account. For that reason this method is only recommended for cases where the surrounding medium has only weak coupling properties, as e.g. air. One can imagine that especially in the presence of water as the load transmitting medium the structural response itself is of great influence on the building-up and decay of the pressure at the structural interface. This is due to the material properties of water, distinguished by low compressibility leading to high pressure rates and cavitation appearing under tension. Therefore the only way true to nature is to solve such problems with a coupled code. Generally it is not obvious to predict whether consideration of coupling effects results in higher or lower load to the structure. However, when great displacements and distortions of the structure are expected, a coupled treatment of the problem is necessary to simulate for instance the afterflow into vacant regions resulting from structural movement.

To get design data for the equipment the calculated accelerations of nodal points within the structure can be interpreted as base excitations of devices situated in the vicinity of them. The approximation to real base excitations depends on the mapping of equipment within the structure. Taking into account the limited capabilities and the high expense of current software and hardware, bolted devices are best modelled as single masses added to the nearest nodal points. Heavy machinery should be represented by elastic bedded rigid bodies able to map the interaction with the foundation by nonlinear springs with damping including all effects of inertia in order to avoid unrealistic influences to the surrounding structure.

The Program

The program family DYSMAS /1/ (**Dy**namic **S**ystem **M**echanics **A**dvanced **S**imulation) has been developed for the simulation of short time physical problems arising mainly in defense. The whole package consists of the two stand alone programs DYSMAS/L, the structure code, and DYSMAS/E, the fluid code, both fully 3-dimensional. A coupling module connects them to the ELC-version.

The scope of application of the structure code covers the analysis of shock loading, vehicle crash, terminal ballistics and missiles. The appendix /L refers to the Lagrangian description of the physical process within a time dependent, material fixed reference frame. The spatial discretization is accomplished by dividing the structure into finite elements of different types like plate-, membrane-, beam-, spring-, volume- and rigid body elements, which can be combined arbitrarily to complex structures. The direct integration of the equations of motion is carried out by means of a central difference scheme /2/. Due to the intense stressing of material during impact, penetration or shock loading the material model covers the entire domain characterized by elastic, nonlinear, failure and post-failure behaviour /3/. A fully incremental formulation is used in the plasticity model, capable of strain hardening. For the representation of arbitrary stress-strain relationships for ductile and brittle materials a sublayer model according to MROZ /4/ is implemented. Beyond this DYSMAS/L contains a lot of further material options, e.g.

a concrete model /5/. A special feature of the L-code is the 3D-multibody contact-processor which is based on a generalized master-slave algorithm. It describes the structural removal within the contact zone between penetrator and target, effects the exchange of momentum and finally determines the contact forces, assuming fully plastic collision locally at the impact area.

Hydrodynamic processes like complex transient flow fields /7/, shock wave propagation in fluids and explosive burn processes are evaluated by the fluid code DYSMAS/E. The Eulerian code uses a space- and time-fixed coordinate system to describe a non-equidistant orthogonal grid. The balance equations for mass, momentum and energy are integrated by an explicit scheme called FLIC (fluid in cell) according to Gentry, Martin, Daly /6/. Equations of state for about 60 different materials such as water, air, soil, concrete including the JWL-equation for explosives and the TILLOTSON-equation are implemented in an intrinsic library. The material strength is described by an elastic/plastic model using the VON MISES yield-condition, isotropic hardening under consideration of strain-rate and temperature dependency. Failure is controlled by instantaneous or cumulative criteria and involves stress tensor modification. Different initiation and burn models for explosives are provided. Particular characteristics of the Eulerian code are the rezoning option to transform the field variables from a 2-dimensional mesh into 3D, or the possibility to mix up to 9 different materials into one cell.

The fully 3-dimensional coupled code DYSMAS/ELC is accomplished by connecting DYSMAS/L and DYSMAS/E by a superposed coupling module DYSMAS/C. Thus a powerful tool was created for the analysis of highly dynamic continuum mechanical problems with special consideration of fluid-structure /8/ and soil-structure interaction phenomena. The physical interaction processes considered, embrace the transfer of loads due to pressure forces and the exchange of energy and momentum accounting for the relative motion of fluid and structure. Reversely, the structural surface, called interface, represents the time-varying boundary condition for the fluid. In this way the fluid-structure interaction code is able to describe the steepening of pressure waves in the water under the retarding influence of the structural interface. On the other hand, the important influence of the fluid-structure interaction is given by the fact, that the pressure load on the structure exerted by the surrounding water is effectively suppressed as soon as the structure interface is moving faster than the neighbouring fluid, for in this case the inertia of the water leads to a breakaway of the flow and the fluid instantly cavitates. A particular attribute is the description of the interface using several simply connected lines in 2D and several simply closed surfaces in 3D. The interface can be arbitrarily oriented relative to the Eulerian grid. This universal formulation permits large movement and deformation of the interface. An extensive possibility of the ELC-code is the definition of complex initial conditions such as pre-stressed structures or initial flow fields, realized by steady-state processors both in the structure and fluid code. An elaborate central time stepping control enables subcycling of each single module depending on the respective grid-sizes and sound velocities. A comprehension of the basic equations for the E- and L-formulations is available in /9/.

The Application

Maritime simulation embraces a wide field of technical problems. The investigation of the response of ships exposed to underwater explosions is a rather outstanding topic primarily appearing in naval defense. The complexity is increasing when treating submarines in comparison to surface vessels, because the pre-stressing of submerged structures by the hydrostatic pressure has to be considered additionally. The code validation for these specific problems was achieved by numerical simulation of different full-scale tests carried out by the German Navy.

The calculation of a torpedo shot against a frigate-sized vessel has been chosen to demonstrate the procedure. Of course, data concerning this topic are to be handled as confidential. Therefore only qualitative results can be presented.

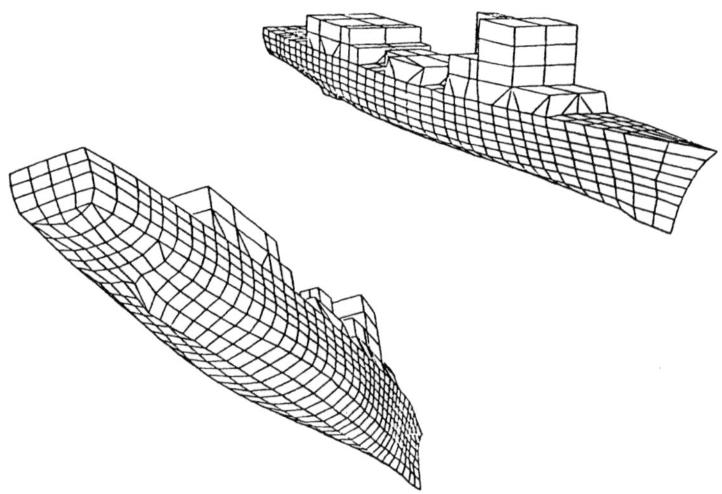

Fig. 1: Finite element model of the frigate

Different views of the finite element model of the target ship are shown in Fig. 1. The contours were digitized from engineering drawings to get the realistic geometric shape. 1764 orthotropic plate elements were generated for the accurate mapping of the stiffness. The additional mass resulting from the equipment was distributed to certain nodal pionts by means of single masses according to design data. The shape of the hull and the mass distribution were checked by evaluating the static equilibrium and the CWL, respectively with the help of a special pre-processor. The adjacent fluid mesh uses about 30000 cells, while the coupling interface is defined by 812 surface elements.

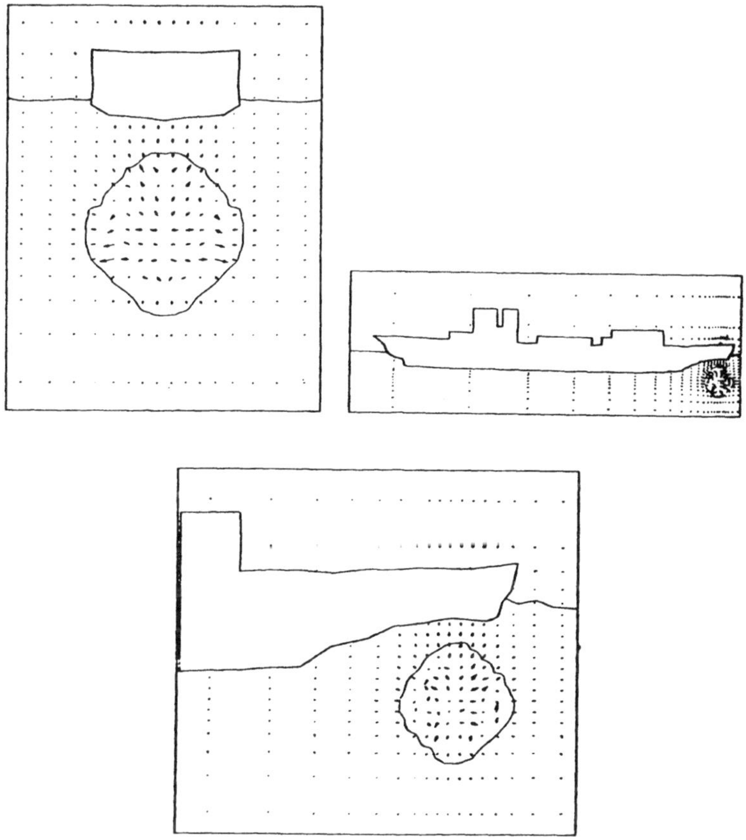

Fig. 2: Vectorplot of velocities containing interface- and material-contours in two perpendicular sections through the centre of the charge in an early state

Fig. 2 illustrates an early state of the gasbubble expansion. Heavy deformations mainly between the bulkheads in the submerged hull regions are evident at this time. The whole stern of the ship is bended upwards like a cantilever beam with varying mass- and stiffness distribution. Local damage in the keel region results primarily from the detonation shock wave arriving immediately after ignition of the charge. In the initial state of this process predominantly inertia forces are acting, because internal forces are built up not till distortions are increasing. That is why the light-weighted hull parts are severely distorted during shock phase. The

The inital spectrum represents the respone during the excitation phase while the residual spectrum is valid for the dying-out phase. Fig. 3 is lining out the response-spectra at four different nodal points marked by symbols in Fig. 4. A four parameter representation is used for the shock-spectrum described by frequency, acceleration, relative displacement and pseudo-velocity. The shock response is decreasing from stern to the centre of the ship by more than one decade approximately proportional to the distance from the admission point. It is obvious that peak values appear at different frequencies in connection with local modes of the structure.

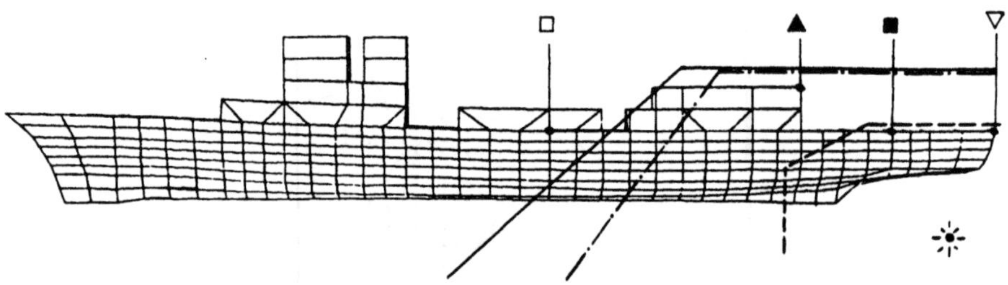

Fig. 4: Damage regions in the ship due to different failure criteria (Symbols refer to Fig. 3)

Shock-response-spectra are best suited for the design of the boltings and elastic beddings of devices and heavy machinery, respectively. Assuming different stability of equipment and evaluating all response-spectra by these critera lead to discrete damage regions for equipment in the ship, as shown in Fig. 4.

The results of such an investigation enables the analyst to make a decision about the residual servicability of the whole system. Further applications of the coupled code up to now are impact of projectiles against water or soil, ground detonations against buildings and different calculations concerning nuclear power plants.

large global displacements of the rear part are growing up over a long time due to the long-term water flow towards the ship caused by the heavy gasbubble expansion. Inside the ship structure the forces are transmitted to the main deck by the stiff bulkheads. The accumulated structural damage reaches severe dimensions.

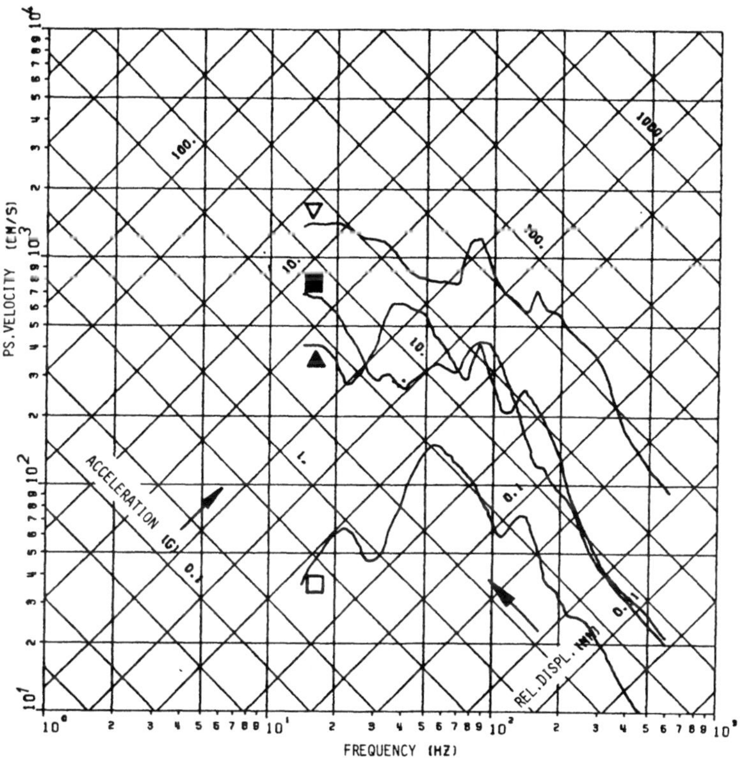

Fig. 3: Shock-response spectra for various nodal points (Symbols refer to Fig. 4)

The load on equipment is connected with the accelations admitting the foundations. However, judgements only based on acceleration transients are not reliable. Therefore the so-called MAXIMAX shock-response-spectra are generated. MAXI MAX stands for a type of response-spectra which takes the maximum response of the initial and of the residual spectrum.

References

1. The Program Family DYSMAS, B-TF-V197, IABG, Munich, 1984

2. Belytschko,T.: Explicit Time Integration of Structure-Mechanical Systems, published in: "Advanced Structural Dynamics", JRC of the Commission of the European Communities, ISPRA, Italy, 1978

3. Poth A., et al.: Failure behaviour of an Aluminium Plate under Impact Loading, Int. Conf. on Appl. of Fract. Mech. to Mat. and Struct., Freiburg, West Germany, 1983

4. Mroz, Z.: On the Description of Anisotropic Work Harding, J. of Mathematics and Physics of solids, Vol. 13, 1957, pp. 163-173

5. Schwarz R., et. al.: Calculation of Impact- and Penetration-Effects on Hardened Reinforced Concrete Structures, Second Symp. on the Interaction of Non-Nuclear Munitions with Structures, University of Florida, Eglin, USA, April 1985

6. Gentry R.A.; Martin R.E.; Daly B.J.: An Eulerian Differencing Method for Unsteady Compressible Flow Problems, Journal of Computational Physics 1, 87-118 (1966)

7. Renn W. et al.: Numerical Simulation of Fuel Spillage Following an Aircraft Impact onto a Nuclear Power Plant, Paper B 3/4, 8th SMIRT, Brussels, Aug. 1985

8. Fießler B. et al.: Use of DYSMAS/ELC for the 3D-Analysis of PWR Fluid-Structure Interaction Problems, Paper B9/8, 7th SMIRT, Chicago, Aug. 1983

9. Zukas, J.A.: Impact Dynamics, John Wiley and sons, New York (1982)

MIX
Papier aus verantwortungsvollen Quellen
Paper from responsible sources
FSC® C105338

If you have any concerns about our products,
you can contact us on
ProductSafety@springernature.com

In case Publisher is established outside the EU,
the EU authorized representative is:
Springer Nature Customer Service Center GmbH
Europaplatz 3, 69115 Heidelberg, Germany

Printed by Libri Plureos GmbH
in Hamburg, Germany